Transactions
of the
American Philosophical Society
Held at Philadelphia
For Promoting Useful Knowledge
Volume 90, Pt. 1

WILLIAM CROONE,
ON THE REASON OF THE MOVEMENT OF
THE MUSCLES

With a Translation by
Paul Maquet

Introduction by
Margaret Nayler

American Philosophical Society
Independence Square ❧ Philadelphia
2000

ISBN:0-87169-901-X
US ISSN:0064-9746

Library of Congress Cataloging-in-Publication Data

Croone, William, 1633-1684.
  [De ratione motus musculorum. Latin]
  William Croone, on the reason of the movement of the muscles /
  with a translation by Paul Maquet ; introduction by Margaret
  Nayler.
      p. cm.-- (Transactions of the American Philosophical Society,
        held at Philadelphia:
      For promoting useful knowledge; v. 90, Pt. 1)
  Includes bibliographical references and index.
  ISBN 0-87169-901-X (pbk.)
    1. Muscle contraction--Early works to 1800. I. Title: Reason of the
  movement of the muscles. II. Maquet, Paul G. J., 1928- III.
  American Philosophical Society. IV. Title. V. Transactions of the
  American Philosophical Society ; v. 90, pt. 1.

  QP321 .C796165 2000
  612.7'41--dc21   00-036247

List of Illustrations

Acknowledgments

I wish to acknowledge the encouragement and support for my work provided by Professor Rod Home of the Department of History and Philosophy of Science in The University of Melbourne, together with the contributions of former staff members of the Department, in particular Mrs. Hazel Maxian and Dr. John Pottage. My PhD thesis, "The Insoluble Problem: Muscle in the Mid to Late Seventeenth Century," has served as a major source for this Introduction. I am also indebted to Dr. Troels Kardel who was instrumental in involving me in this project and encouraging my participation.

I also acknowledge the assistance provided by the Library Staff of the Royal Society when I was researching Croone among other seventeenth-century writers on muscle contraction. Lastly, I wish to thank the President and Council of the Royal Society for permission to publish the sketch of Croone's bladder and strings experiment from the Register Book Original.

Margaret Nayler

Troels Kardel, M.D., sent me a copy of the 1680 edition (anonymous) of the *Ratione Motus Musculorum* and suggested that I translate it and ask Margaret Nayler for the erudite introduction which she kindly provided.

I checked with first edition of 1664, also anonymous, a microfilm of which was kindly provided by the Wellcome Trust in London at the request of the Furlong Research Foundation. The following Latin text is a print of this microfilm.

Paul Maquet, M.D.

BRIEF BIOGRAPHY [1]

William Croone (1633-1684)

William Croone was born in London and educated at the Merchant Taylor School and at Emmanuel College, Cambridge, where he took his first degree in arts in 1650; he was elected to a fellowship in 1651. In 1659 he held the position of Professor of Rhetoric at Gresham College, London, where his colleagues included Wren, Petty and Goddard. When the Royal Society was formed, he was appointed Register, being succeeded by John Wilkins and Henry Oldenburg when the charter was passed. In 1662 Croone was made a doctor of medicine by royal mandate. In May, 1663, he was chosen as one of the first fellows of the Royal Society, and remained an active member, frequently serving on the Council, until his death. He succeeded Sir Charles Scarburgh as lecturer in anatomy at Surgeons-hall in 1670, and in the same year resigned his Gresham College professorship prior to his marriage to the daughter of Sir John Lorimer. Croone conducted a medical practice in the City of London, and was admitted a fellow of the College of Physicians in 1675. He died of a fever in October, 1684.

Apart from his work on muscles, Croone displayed an interest in numerous and varied subjects. His biological work within the Royal Society covered aspects of respiration, circulation, skin-grafting, blood transfusions and embryology; in the physical sciences he investigated freezing solutions, air density, the breaking of wires and phosphorescence.

In his funeral sermon he was described as "not only a friend, but an ornament to the whole race of mankind: a general scholar, an accurate linguist, an acute mathematician, a well-read historian and a profound philosopher; eminent for his generosity and charity; amiable in his temper, prudent in his conduct, chearful and facetious in his conversation and possessed of a just sense of the duties of religion."[2]

Croone left a plan for two lectures, one to be given at the College of Physicians on the nerves and the brain, and the other to be given at the Royal Society upon the nature and laws of muscular motion, accompanied by some anatomical demonstration. Unfortunately his will made no provision for the endowment of these lectures, and funds were not available until 1701 by the terms of his widow's will. The first Croonian Royal Society lecture was given in 1738 by Alexander Stuart: the first College of Physicians lecture was given by Thomas Lawrence in 1749.[3]

---

[1] This is based on L. M. Payne, L. G. Wilson and Sir H. Hartley, "William Croone, F.R.S. (1633-1684)," *Notes and Records of the Royal Society of London*, 15 (1960): 211-19.

[2] Thomas Birch, *The History of the Royal Society*, (London: 1756-7; facsimile edition, New York: 1968, Sources of Science 44), Vol. 1V, pp. 339-40.

[3] A full discussion of the Croonian Lecture Fund is given in *The Record of the Royal Society of London*, (London: Oxford University Press, 3rd ed., 1912), p. 176.

INTRODUCTION

When William Croone[4] published his small treatise, *De ratione motus musculorum* in 1664, it represented one of the earliest attempts to explicate muscle contraction in terms of the then current mechanical and chemical concepts. The work is significant not only because it provides an informative overview of the difficulties inherent in addressing the question of how muscles contract, but also because it derives from a series of experiments that form a logical framework for the notion that expanding muscle, like a bladder filled with air or water, can exert a force capable of moving parts of the body against considerable resistance. Croone also adopted the attitude to chemistry aptly expressed by his Cambridge contemporary and later fellow Royal Society member, Henry Power, for whom

> all the operations of nature within us are most emphatically expressed and indeed repracticed by the chymists, without us, and therefore the great and mysterious works of concoction, chylification, sanguification, assimilation etc. are the most powerfully demonstrated by Chymicall Analogy.[5]

Such a belief underpinned Croone's proposal that it was an expansive chemical reaction between blood and nervous fluid that caused the muscle to swell and thus contract.

There is no doubt that muscle presented a unique physiological problem in the seventeenth century and earlier, insofar as its main function, contraction, was an established fact in contrast to other organs,

---

[4] I have adopted the spelling "Croone" as printed on the title page of the 1667 edition of *De ratione motus musculorum* and as preferred by William Munk in *The Roll of the Royal College of Physicians* (London: 2nd ed., 1878). Thomas Birch in *The History of the Royal Society* adopts "Croune," which is the spelling that appears on his tombstone, and is the commonest alternative spelling found in secondary sources. A footnote to Croone's obituary, printed in Birch (Vol. IV, p. 339) refers to another four variations found in printed books. The Royal Society register, which members signed on attending each meeting, shows that Croone himself adopted a number of variations and was certainly more inconsistent than most of his fellow signatories.

[5] Henry Power, "Analogia Physico-Chymica," May, 1657, from Sloane MS 1393, f7, p. 46, as quoted by A. B. Davis, *Circulation Physiology and Medical Chemistry in England, 1650-1680*, (Kansas: Coronado Press, 1973), p. 23.

such as the liver and the pancreas, whose functions were by no means obvious. As Walter Charleton commented, in 1659, "undoubtedly, the theory and function of this [muscle] is as disputable and obscure to the intellect as the phenomenon itself is apparent to the senses."[6]

BACKGROUND IDEAS

Prior to the upsurge in ideas that occurred after 1650, conceptions about how muscle contracted were generally imprecise and couched in qualitative terms. John Bulwer, writing in 1649, justified his discussion of the muscles of facial expression and emotions with the premise that it is a "disparagement" to the educated man "to be as a meere Puppet or Mathematicall motion, and not to understand why, or after what manner, the Muscles of his Head move in obedience to the command of his will. . . ."[7] But his explanation of muscle as an instrument to the motive faculty or the animal spirits, receiving the command to move from the brain via the nerve and responding because of an "ingenit virtue"[8] does not seem to be very enlightening for the reader. Bulwer does raise the issue of the speed with which muscles move in response to the will which mitigates against a time-consuming flow of material spirit from brain to muscle. By way of explanation he proposes that the motive faculty "doth perpetually flow and travel to the Nerves"[9] and thence to the muscle so that the idea of moving and the resultant movement are inseparable: motive faculty seems to defy further description. Similar statements with respect to the nerve "carrying the faculty of moving" and muscle being "prepared and fitted to

---

[6] Walter Charleton, *Oeconomia animalis: novis in medicina hypothesibus superstructa et mechanice explicata.* Charleton acknowledges his debt to Scaliger who first made this comment in relation to light and colors. The abbreviated English version of this work, published in the same year, 1659, and titled *Natural History of Nutrition, Life and Voluntary Motion*, does not contain this comment.

[7] John Bulwer, *Pathomyotomia or a Dissection of the Significative Muscles of the Affections of the Minde, Being an essay to a new method of observing the most important movings of the muscles of the head . . .* , (London: 1649), p. 2. An entry in Hooke's diary records that among the books he purchased in 1678 was "Bulwer of Muscles." See Henry W. Robinson and Walter Adams, *The Diary of Robert Hooke, M.A. M.D. F.R.S. 1672-1680,* (London: Taylor and Francis, 1935), p. 362.

[8] Ibid., p. 28.

[9] Ibid., p. 22.

receive the influence of the moving quality" are to be found in Helkiah Crooke's all-embracing *Microcosmographia*.[10]

To some extent this imprecision in relation to muscle activity stemmed from earlier writers. Galen had said that muscle consistently exhibited the property of innate contractility; in voluntary or willed movement one of a pair of muscles pulling equally on either side of a joint contracted more strongly to overcome its antagonist and produce movement. The directive for this stronger contraction came from the brain via the nerve, but Galen was undecided as to whether this was effected by the direct flow or excitation of the psychic pneuma [animal spirits], by the transmission of some quality through the pneuma, or even by the nerve mechanically pulling on the muscle.[11] The muscle structure Galen described, with the nervous fibers blending with binding tissue from the head of the muscle to form tendinous fibers at the end of the muscle[12] certainly did not lend itself to shortening brought about by being filled with spirit.

Fabricius maintained that tendon surrounded by flesh in the belly of the muscle was the contractile element: nerve activated the inherent contractile faculty by serving as the passive transmitter of a quality like light or magnetism from brain to muscle without requiring direct contact for action.[13] Fabricius dismissed the idea of muscle being filled with spirit, and followed Galen in claiming that "not only the whole muscle but also every particle of it has innate motion, namely contraction into itself."[14]

---

[10] Helkiah Crooke, *Microcosmographia, A Description of the Body of Man Together with the Controversies and Figures thereto belonging*, (London: Thomas & Richard Coles, 2nd ed., 1631, pp. 740-741. A similar statement appears in the 1651 ed., p. 554.

[11] This is fully discussed in M. A. Nayler, "A Thorny Problem: Galen, Fabricius and Harvey on Muscle," unpublished M.A. thesis, University of Melbourne, 1975, Chapter II, pp. 79ff.

[12] Galen is often incorrectly quoted as saying that nervous fibers blended with ligamentous fibers from the head of the muscle. Goss has rightly pointed out that the Renaissance scholars translated the Greek 'syndesmos' as 'ligament' rather that binding substance, thus leading to some confusion when 'ligament' came to be primarily associated with tissue binding bone to bone. See Galen, "De motu musculorum" translated by C. M. Goss in "On Movement of Muscles by Galen of Pergamon," *American Journal of Anatomy*, 123 (1968): 24-25.

[13] Nayler, "A Thorny Problem," Chapter IV, pp. 143ff.

[14] Ibid., p. 62. (See Hieronymous Fabricius of Aquapendente, "De musculi actione" [1614], one of the tracts comprising *De musculi artificio*, [Padua: 1625], p. 88).

Harvey attributed muscle contraction to the shortening and hardening of the fleshy fibers, dependent in some undefined way on the motive spirit, an intrinsic part of the muscle derived from blood and ultimately the heart, rather than nerve from the brain that was primarily responsible for regulating muscle movements.[15] The rough notes for Harvey's intended work on muscle, *De motu locali animalium* (commenced in 1627),[16] are largely summaries of earlier opinions, and Whitteridge has suggested that it may have been Harvey's realization that "he was confronted with a series of questions which he had no hope of answering" which prevented him from pursuing his work on muscle.[17]

Nevertheless, Harvey's contemporaries would have been in no doubt about his abhorrence of spirits as an explanatory mechanism: never demonstrated by dissection, they "serve as the common subterfuge of ignorance."[18] Harvey was also unimpressed by mechanical analogies, although his work on the circulation could be seen as illustrative of a unique mechanistic and mathematical mode of thought. Arguments against this view have pointed out that Harvey's calculations on the quantity of the blood passing through the heart were "of the simplest possible nature":[19] they were corroborative rather than fundamental to his elucidation of the circulation of the blood. Unable to explain the movement of the heart beyond its being muscular and initially set in motion by the expanding blood, Harvey was scathing of any thought that it operated as a heating device to vaporize blood, as Descartes had

---

[15] Nayler, "A Thorny Problem," Chapter V, pp. 194ff.

[16] William Harvey, *De motu locali animalium 1627*, edited and translated by Gweneth Whitteridge, (Cambridge: University Press, 1959), pp. 141-143.

[17] Gweneth Whitteridge, "Of the Local Movement of Animals: the Wilkins Lecture, 1979," *Notes and Records of the Royal Society of London*, 34 (2) (1980): 142.

[18] Harvey, "A Second Disquisition to Jean Riolan," 1649, in *The Works of William Harvey, M.D.*, translated by Robert Willis, (London: Sydenham Society, 1847; reproduced in Sources of Science No. 13, Johnson Reprint Corporation, New York, 1965), p. 116.

[19] G. Whitteridge, "Of the local Movement of Animals . . . ," p. 145; also Theodore M. Brown, "The Mechanical Philosophy and the 'Animal Oeconomy'; A Study in the Development of English Physiology in the Seventeenth and Eighteenth Century," Ph.D. thesis, Princeton University, (Ann Arbor, Michigan: University Microfilms Inc. 1968), pp. 13-22.

proposed.[20] As late as 1652 Harvey wrote that the two movements of contraction and relaxation of the heart muscle "inhere in the substance of the heart itself, just as they do in all other muscles."[21] Beyond this, Harvey never attempted any explanation of how the muscle fiber might contract, although his notes reveal that he was well aware of the proposed mechanical analogies linking shortening and distension with the addition of fluid as in wet rope, or some sort of expansion of the fluid content (i.e. ebullition).[22]

Croone could thus be said to have inherited an approach that concerned itself with careful observation and description without recourse to theorizing as to how muscle contracted. Yet he chose to look for explanations in terms of muscle being quantitatively distended and shortened, rather than to defer to an unexplained motive faculty, largely because the mechanical conception of nature seemed to offer a more promising approach.

This mechanical view of nature developed from a number of strands of thought, including Galileo's premise that nothing is scientifically knowable except what is measurable, Gassendi's revival of atomism, and Descartes' conviction that all natural phenomena were to be explained in terms of particulate matter in motion and subject to mechanical laws.[23] Robert Boyle, whose views were well known in both England and the Continent, developed his own corpuscular philosophy, declaring that "men do so easily understand one another's meaning when they talk of local motion, rest, bigness, shape, order, situation and contexture of

---

[20] Harvey, "A Second Disquisition to Jean Riolan," p. 137.

[21] Harvey, Letter to R. Morrison, M.D., of Paris, 28th April, 1652, *The Works of William Harvey*, p. 604.

[22] Harvey, *De motu locali animalium*, p. 101.

[23] The mechanical philosophy has been discussed in detail by a number of authors and from varying viewpoints, in particular: R. G. Collingwood, *The Idea of Nature*, (Oxford: Clarendon Press, 1945); Marie Boas, "The Establishment of the Mechanical Philosophy," *Osiris*, 10 (1952): 412-541; E. J. Dijksterhuis, *The Mechanization of the World Picture*, translated by C. Dikshoorn, (Oxford: Clarendon Press, 1961); Richard S. Westfall, *Science and Religion in Seventeenth Century England*, (Ann Arbor: Michigan University Press, 1973, being paperback edition of 1958 publication); and Richard S. Westfall, *The Construction of Modern Science: Mechanisms and Mechanics* (Cambridge: University Press, 1977, first published by John Wiley, 1971).

material substances."[24]

Croone would certainly have been familiar with all these views, together with Regius's (1646)[25] specific explanations for muscle contraction, also those of Descartes (1634 but not published until 1662)[26] and Walter Charleton.[27] The systems proposed by Regius and Descartes paid little attention to muscle structure, but concentrated on the mechanisms necessary to regulate the inflationary spirit flow from nerve to muscle, and between oppositely acting muscles. Given that visible valves controlled the flow of materials in parts of the body such as the heart, gall bladder and veins, it was seen to be logical to assume that invisible valves similarly controlled the flow of the animal spirits that inflated the muscles, just as air inflates and hardens a ball. While their "Tubes, Valves and Openings" were rightly judged to be "Children of the Imagination,"[28] Descartes and Regius did focus attention on certain aspects of muscle contraction, namely:

1. the speed of the action, which is not consistent with the time taken for a fluid (animal spirit) flow from brain to muscle.

2. the quantity of animal spirit required, again not consistent with the brain acting as a total storage unit, and thus necessitating flow from the antagonist muscle.

3. the nature of the animal spirits as the most subtle, lively and strong particles of the blood and capable of exerting sufficient strength to inflate a muscle.

As both Croone and Charleton were early and active members of the Royal Society, which began its meetings in December 1660, the opportunity for discussion on a subject of mutual interest must surely

---

[24]Robert Boyle, "Of the Excellency and Grounds of the Corpuscular or Mechanical Philosophy," 1674, *The Works of the Honourable Robert Boyle*, printed for A. Millar in 5 vols, (London: 1744), Vol. III, p. 451. This essay was written in 1665, but Boyle's views would have been well-known prior to this date.

[25] Henricus Regius, *Fundamenta physicis*, (Amstelodami, 1646), Chapters X and XII. Repeated in *Philosophia naturalis* 1654 and 1661.

[26] René Descartes, *De homine figuris, et latinate donatus a Florentia Schuyl*, (Lugduni Batavorum), 1662.

[27] Charleton, *Oeconomia animalis . . .*, Chapter 11, and *Natural History . . .*, Chapter 11.

[28] James Parsons, "The Croonian Lectures of Muscular Motion, 1744 and 1745," *Philosophical Transactions*, Supplement to Vol. XLIII, p. 8.

have presented itself.[29] Yet there is no evidence to suggest that Croone was in any way influenced by Charleton's anachronistic geometrical argument that a non-contracted muscle (eg. Biceps) represented by a parallelogram is converted into a square of the same dimensions when it contracts. Constancy of volume is implied, but not demonstrated, and in conflict with contraction effected by the influx of animal spirit.[30] Perhaps Charleton's most significant and influential observation is as follows:

> For, it seems more reasonable, that the swelling in the body of the Muscle is the Cause of its Contraction; than, on the contrary, that the Contraction should be the cause of the Swelling, as those contend who would have the motion to be performed without the afflux of spirits.[31]

THE EXPERIMENTAL FRAMEWORK FOR
DE RATIONE MOTUS MUSCULORUM

As temporary Registrar to the newly formed Royal Society, Croone would have been well aware of its explicit stress on the importance of experiments: in the words of the first Charter granted in July, 1662,

> we look with favour upon all forms of learning, but with particular grace we encourage philosophical studies, especially those that by actual experiments attempt either to shape out a new philosophy or to perfect the old.[32]

In the light of this emphasis it is, then, surprising that the clearest statement of the experimental basis for Croone's 1664 hypothesis is to be

---

[29] Thomas Birch, *The History of the Royal Society*, (London: 1756-7; facsimile edition, New York: 1968, Sources of Science 44), Vol. I, pp. 3-7. Charles Webster, *The Great Instauration: Science, Medicine and Reform, 1626-1660*, London: Duckworth, 1975, Table 1, p. 92.

[30] Charleton, *Oeconomia animalis . . .* ,pp. 287-289; *Natural History . . .* , pp. 207-208. Charleton gives Jacob Muller, a pupil of Gregor Horst, as his source (in *Oeconomia animalis . . .* only).

[31] Charleton, *Natural History . . .* , p. 188.

[32] *The Record of the Royal Society*, 3rd ed., (London: Oxford University Press, 1912), p. 59. See also Michael Hunter, *Science and Society in Restoration England*, (Cambridge: Cambridge University Press, 1982), p. 37.

found in a later paper where he affirms that this earlier hypothesis was "grounded on an Experiment made for that purpose before the Royal Society a little before that time."[33] The experiment referred to is entered in the Register Book of the Royal Society as "An Experimentall Account of the raising up of a weight hung at the bottome of an emptie Bladder," and was performed on September 4th, 1661.[34] Some four weeks earlier John Wilkins had completed a series of experiments "concerning the force of blowing with a man's breath."[35] Croone refers to both Wilkins's experiments and his own in the final paragraph of his 1664 treatise where he states that heavy weights attached to a bladder are elevated rapidly enough by inflating the bladder with air or water. This would seem to confirm that these experiments played a significant part in the development of Croone's ideas.

Wilkins's experiments, performed at the Royal Society meetings on July 31st and August 7th, 1661, were apparently designed to demonstrate that air blown into a bladder could exert sufficient force to lift weights. The first experiment demonstrated that the bladder of an ox or cow could, when inflated, raise a heavy weight some two inches (see Figure 1A). In the second experiment, a weight was attached to one end of the bladder (Figure 1B); when the bladder was inflated, the weight was raised some three to four inches off the ground. The third experiment consisted of enclosing the bladder in a box, placing a weight of one hundred and ten pounds or more on a flat board on top of the bladder, and then inflating it (see Figure 1C). There is no discussion recorded concerning these experiments, in accordance with the statutes whereby "the matter of fact shall barely stated, without any prefaces, apologies or rhetorical flourishes; and entered so in the Register-book"; any conjectures "concerning the cause of the phenomena in such Experiments, the same shall be done apart" and entered into the Register-book only if ordered.[36]

---

[33] Croone, "An Hypothesis of the Structure of a Muscle and the reason of its Contraction: read in the Surgeon's Theatre Anno 1674, 1675." *Philosophical Collections*, 1681, p. 25.

[34] This account is in the Register Book (Original) of the Royal Society, Vol. I. (1661-1662), pp. 108-112. From the entry in Birch, Vol. I, p. 42, it would seem that the experiment may not have been successful at the earlier meeting on August 28th.

[35] Birch, *The History of the Royal Society*, p. 36.

[36] *The Record of the Royal Society of London*, p. 119. These Statutes were enacted in 1663, but were clearly in operation as regards the Register Book from its inception.

According to Frank, experiments with bladders raising weights were first undertaken at the Oxford Experimental Philosophy Club in 1650, and described in a letter from Petty to Hartlib dated December 16th, 1650:[37] this reference does not clarify whether all three of Wilkins's 1661 experiments were identical with the earlier ones. In reference to linking these experiments to muscle contraction by inflation both Bathurst (1654)[38] and Willis (1661)[39] made this inference—Willis recording that he had seen 150 pounds of lead raised by a bladder inflated with air. It is conceivable that the repetition of these experiments may have been prompted by experiments that Robert Boyle had performed and reported in *New Experiments Physico-Mechanical* . . . , 1660. Boyle had enclosed a lamb's bladder, half full of air, in his receiver and noted that when the surrounding air was removed the bladder swelled, and sometimes the air in the bladder was seen to "suddenly expand itself so much and so briskly that it manifestly lifted up some light bodies that leaned on it, and seemed to lift up the bladder itself."[40]

Wilkins's experiments seem to offer confirmation that air could exert a comparatively large force if a sufficient quantity is confined within an expandible, minimally elastic container such as a dried bladder. This force could raise weights vertically, and the vertical expansion could also effect a horizontal shortening capable of raising an attached weight as in the second experiment. The idea of enclosing the weighted bladder in a box was possibly to observe the supposed effect of lessening the weight of the atmosphere, presuming that the box, "open at one end," was closed during the experiment. However, no comparison was drawn between the weight lifted within the box and the unspecified weight of the boy lifted in the first experiment. Experiments with bladders performed at the Accademia

---

[37] Robert G. Frank Jr., *Harvey and the Oxford Physiologists,* (Berkeley: University of California Press, 1980), p. 55.

[38] Ibid., p. 108, refers to Ralph Bathurst's *Praelectiones tres de respiratione,* 1654; unpublished but read by Willis, Boyle and Lower.

[39] Kenneth Dewhurst, *Thomas Willis' Oxford Lectures,* (Oxford: Sandford Publications, 1980), p. 55. Lower's notes from Willis's 1661 lectures were sent, on request, to Boyle in September, 1662.

[40] Boyle, "New Experiments Physico-Mechanical Touching the Spring of the Air, and its Effects, Made for the Most Part in a New Pneumatical Engine," December 20th, 1659, in *Works* . . . , Vol. I, pp. 12-13. This work was published in 1660.

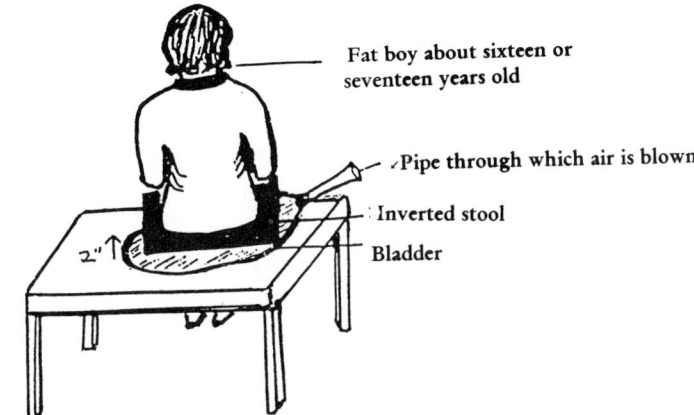

Fat boy about sixteen or seventeen years old

Pipe through which air is blown

Inverted stool

Bladder

2"

Figure 1A. Wilkins's first experiment

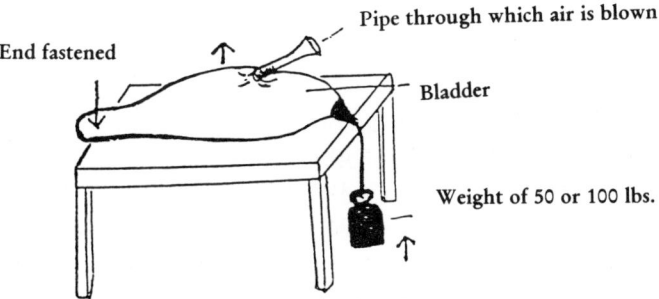

Pipe through which air is blown

End fastened

Bladder

Weight of 50 or 100 lbs.

Figure 1B. Wilkins's second experiment

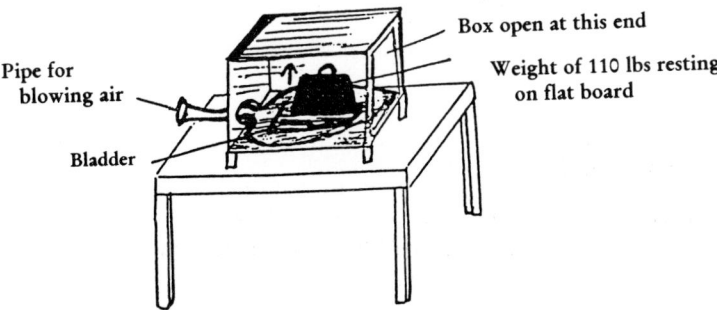

Box open at this end

Pipe for blowing air

Weight of 110 lbs resting on flat board

Bladder

Figure 1C. Wilkins's third experiment

Figure 1. Wilkins's experiments with inflated bladders as interpreted by the author.

del Cimento involved enclosing the whole apparatus in a bell jar to counter the effects of atmospheric pressure, but it is unlikely that the Royal Society was aware of the details of this work.[41]

Croone would seem to have been impressed by the seemingly small effort exerted in one direction which could effectively overcome a considerable resistance acting at right angles to it (as exemplified by Wilkins's second experiment), and set out to illustrate this more clearly. In his 1661 experiment (see Figure 2) a bladder AEHD was first suspended full of water to stretch it out; it was then emptied and a weight, B, of 7lb attached to its lower end so that it just touched the line ZY. "Two pounds or pints of water" were poured into the bladder, and it was noted that the weight rose "somewhat more than an inch." Next, in an attempt to simulate the "forces" acting, two strings representing the walls of the bladder were fastened together at A and H, with the same weight as before being attached at H with its lower end touching ZY. To illustrate the horizontal action of the water filling the bladder, two 1lb weights were attached to the strings at D and E respectively, the result being that "the seven pound weight was raised to the same height above ZY as before." The remainder of Croone's paper is devoted to a geometrical analysis of the "forces" or "pressures" acting to produce this change.

Croone contends that if the quadrilateral ARCQ represents the bladder filled to the level RPQ, and AEHD represents the bladder filled to the higher level EFD, then as the weight at C rises to H, the angle RCQ is increased to EHD. Croone continues that

> to explicate this dilatation of the angle RCQ so as to accommodate it to the Experiment of the Strings, I can find no other way (I humbly submitt it to others better judgement) then by assumeing for an Hypothesis, that the water contain'd in the Triangle RCQ has a pressure every way . . .

for example,

> the water that is in the point P, it has not only an action

---

[41] Details of these experiments are to be found in W.E.K. Middleton, *The Experimenters. A Study of the Accademia del Cimento*, (Baltimore: Johns Hopkins University Press, 1971), pp. 111-113. The experimental work of the Accademia del Cimento was not published until 1667, an account being given at the meeting of the Royal Society on March 19th, 1667/8. However, the experiments on pneumatics had been largely completed prior to 1660, 1657 being given for the particular experiments referred to in the above text.

downward perpendicular uppon the point C, but also uppon every point of the side CQ, according to all right lines that can possibly be drawn from P between C and Q although I deny not, but its greatest action or pressure may be the perpendicular.

Accepting for the sake of argument Croone's supposition that the water exerts both a horizontal pressure along the lines PQ and FD, and a vertical pressure at C and H, then the added pressure at Q can be equated with the 1lb weight at G, and the added vertical pressure can be equated with the 7lb weight at C. Given this, Croone adds that "it will bee easy to show that . . . the least lateral pressure of the water against the sides of the bladder . . . shall overcome the greatest perpendicular pressure," and this is proved, in part, by the experiment of the strings where the least inflexion of the extended string ADH would cause the weight C to rise above the line XZ.

By further geometrical construction, Croone proves that since AD + DH = AB (same piece of string, AB being the length when the string is vertical and just touching ZY), and AF = Aω and HF = HQ (radii of quadrants centered on A and H respectively) then the remainder ωD + QD = the remainder HB; thus FD is to ωD + QD as the weight G is to the weight C. Galileo is cited as having demonstrated that the proportion of the tangent FD to ωD +QD, or singly of either ωD, or QD [the horizontal displacement to the vertical displacement] "shall always be greater than the greatest weight at H to the least at G."[42] Lastly, the further D recedes from F, the greater is "the resistance in B" in relation to "the force in G."

Croone concludes that

> the same cause appears in the bending of bowes and the raising

---

[42] Croone's reference is to Galileo's "Dial; di movement: local " suggesting *Two New Sciences*, but Day 3 on "Local Motion" deals with acceleration. I can only suggest that the reference is to Galileo's dialogue version of his essay 'De motu,' this and the essay version being tentatively dated by Drabkin as between 1589 and 1592. Neither was published by Galileo, but possibly Croone had access to a manuscript copy. The most relevant passage that I can find, in the essay version, is in relation to inclined planes where Galileo says that "the same weight can be drawn up an inclined plane with less force than vertically, in proportion as the vertical ascent is smaller that the oblique." See Galileo Galilei, *On Motion and On Mechanics*, translated by I. E Drabkin and Stillman Drake,(Madison: The University of Wisconsin Press, 1960), p. 65.

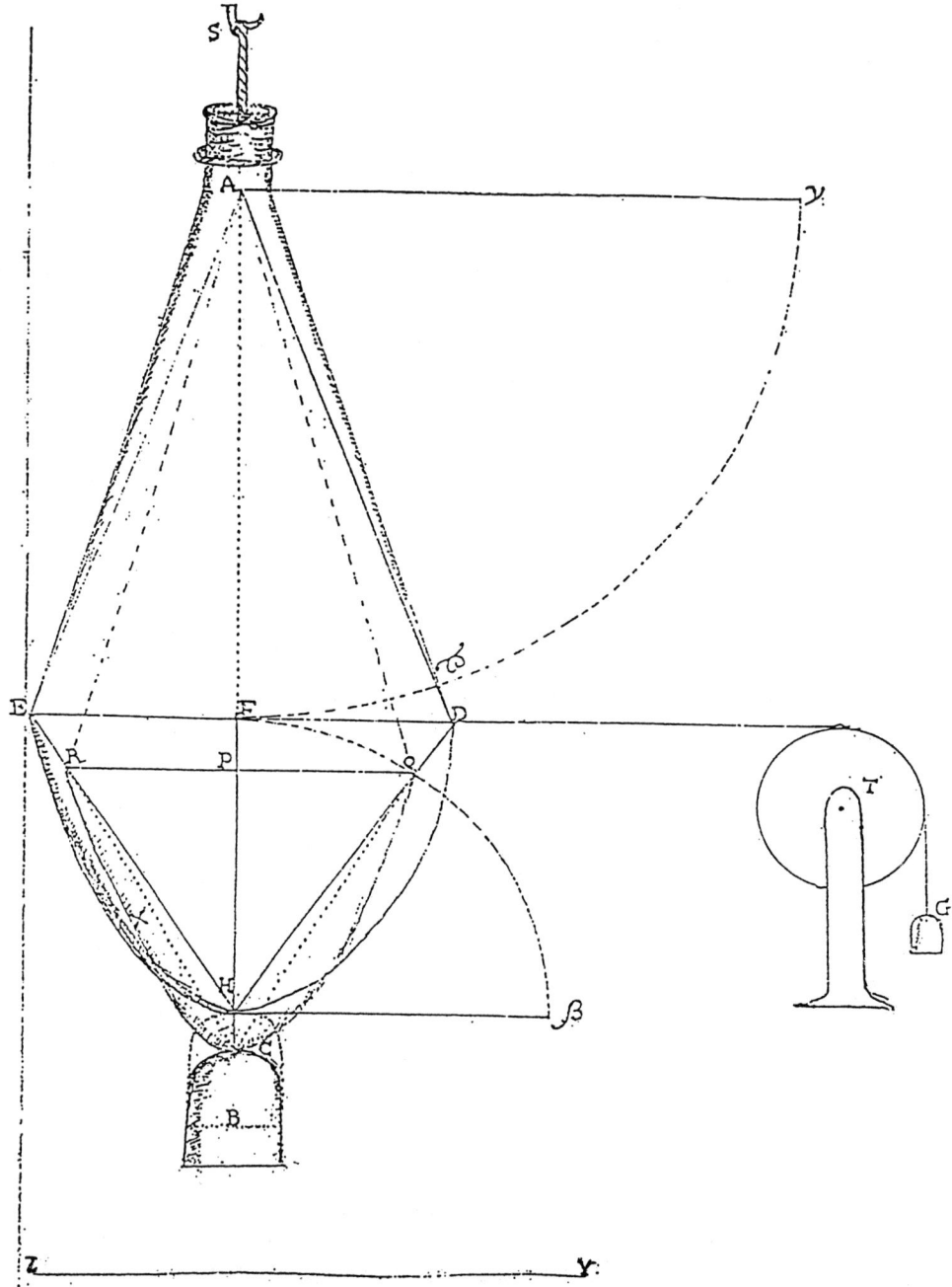

Figure 2. Croone's experiment with bladder, weights and strings (from Register Book [original])of the Royal Society, Vol.1.,1660-1662, p. 108. Note that only one of the two horizontal weights acting at D and E is included in this illustration.

of great weights by wetting ropes, which is but a greater multitude of distensions and contractions in the small parts, and seems a very agreeable Hypothesis, to explicate the Doctrine of Muscular Motion.[43]

The raising of weights by wetting ropes is an analogy drawn by Fabricius when discussing the nature of muscle contraction.[44] Galileo deals briefly with wet rope in his *Two New Sciences* (1638) where the First day discussions center on the resistance of solid bodies to separation; the force exerted by wet rope can be explained by supposing that "innumerable atoms of water" are driven between its fibers when it is exposed to wind-blown fog. The additive effect of all these very small expansions that shorten the rope is a large force capable of moving a very heavy weight.[45] Croone offers no further information as to how the wet rope analogy influenced his thinking, nor how it might be applied to a muscle comprised of numerous fibers like strands of rope.

It is difficult to conclude, from Croone's account, that he actually conducted his experiments for the first time in the sequence described. Either the lengths of string would have had to be measured beforehand, or else the weights pulling horizontally predetermined, in order to produce an elevation of the 7-lb weight exactly equivalent to that produced when the bladder was filled with water. In reference to his discussion, it is evident that Croone did not understand the behavior of fluids and hence he assumes a lateral pressure acting at the surface of the water where the pressure is actually zero. His confusion over forces and pressures makes interpretation difficult. Nevertheless, if the bladder is seen as a non-elastic container, the analogy of the strings is reasonable despite its inaccuracies regarding the behavior of fluids. As an isolated experiment—with only a hint as to its application to muscular motion—its significance is only apparent in the light of subsequent development.

---

[43] Croone, An Experimentall Account . . . , pp 108-12.

[44] Fabricius, *De musculi actione*, p. 87.

[45] Galileo, *Dialogue Concerning Two New Sciences*, translated by Henry Grew and Alfonso de Salvio, (New York: Dover Publications, 1914), pp. 19-20. Croone would have been familiar with the account of the erection of the Vatican obelisk in St Peter's Square, Rome, by Fontana in 1586; only by wetting and thus shortening the ropes could the obelisk be successfully placed in position. See *Archives Internationales d'Histoire des Sciences*, 1949, Vol. 28, pp. 827ff.

According to Wilson, "it was this experiment that indicated to Croone that an internal force of expansion, causing a muscle to swell, would also cause it to shorten and which was therefore, one of the foundations of his theory"[46] as developed in his 1664 paper. This same concept follows from Wilkins's second experiment, and is basic to earlier ideas of muscle contraction by inflation. The significance of Croone's experiment is the attempt to measure the forces involved, and thus to demonstrate that the internal force of expansion need not be very great in order to effect a shortening against considerable resistance.

Raising weights by small forces remained a topic of interest at the Royal Society for some time following Wilkins's and Croone's experiments, and on March 4th, 1662/3, John Wallis brought in his analysis of Wilkins's second experiment.[47] Based on static principles, Wallis correlates the diameter of the pipe through which air is blown from the lungs with the altitude of the bladder that is regarded, in the first instance and for convenience, as a rhombus. These measures are related to the "weight" of the incoming air and the weight to be raised. Wallis calculates that if the weight and strength of the incoming air is equivalent to 1/5000th of the weight to be raised, then it will lift it through a distance that is 1/12th of the length of the bladder. In justifying the use of a straight rather than a curved figure for his detailed calculations, Wallis notes that this is near enough "for the present purpose, the wondrousness of this experiment not arising so much from the particular forme of the Bladder, as from the Ascent of soo great a weight by soo small a force." The "force" referred to is that of the lungs which in turn is equated with the strength of the muscles of the thorax.

The production of movement by the absorption of moisture was the subject of investigation at Royal Society meetings in 1663 and 1664, with both Goddard and Hooke developing hygroscopes, the former using lute strings,[48] and the latter using the beard of the wild oat and later, the pod

---

[46] L.G. Wilson,. "William Croone's Theory of Muscular Contraction," *Notes and Records of the Royal Society of London*, 16 (1960): p. 164.

[47] John Wallis, "An Account of the Experiment wherein a Weight is raised by the Blowing of a Bladder," Register Book (Original) of the Royal Society, Vol. II, pp. 120-137.

[48] Goddard's idea of a hygroscope with pulleys and lute strings was proposed at the Royal Society meeting of October 7th, 1663 (Birch, *The History of the Royal Society*, Vol. I, p. 311), and was doubtless a development of earlier work on gut string shrinkage: at the meeting of July 6th Goddard had promised to demonstrate a "way of raising a considerable

of the vetch.[49] Hooke quite specifically linked fluid absorption by plants to explaining animal motion. In his *Micrographia* (1665)he states that "were this Principle very well examin'd I am very apt to think it would afford us a very great help to find out the Mechanism of the Muscles" for just as the wild oat beard held near the fire straightens when its "knee" or angle is moistened with "well rectify'd spirit of wine" and then bends as the wine evaporates, "so may, perhaps, the shrinking and relaxing of the muscles be by the influx and evaporating of some kind of liquor or juice."[50] Doubtless Hooke had expressed these ideas in the discussions of his experiments long before they appeared in print, and Croone would have been familiar with them.

DE RATIONE MOTUS MUSCULORUM

On May 4, 1664, it is recorded that Dr Croone presented his little treatise, *De ratione motus musculorum*, to the Royal Society "by the hands of Mr Hill."[51] This first edition was published anonymously in London, and was later to cause some confusion as a result of being appended to Thomas Willis's *Cerebri anatome*, 1664.[52]

The plan of exposition that Croone attempts to follow, and which is discernable in spite of numerous digressions, can be summarized as follows:

• Earlier opinions on the cause of muscle movement.
•The presuppositions, both anatomical and physiological,

---

weight by shrinkage of gut strings," although there is no record of this experiment being carried out (Birch, *The History of the Royal Society*, Vol. I, p. 271).

[49] Birch, *The History of the Royal Society*, Vol. I, p. 320; Vol. II, p. 100.

[50] Robert Hooke, *Micrographia*, 1665. Reprinted in *Early Science at Oxford*, R. T. Gunther(Ed), Vol.XIII (Oxford: University Press, 1938), pp. 151-152. An illustration of the hygroscope is found in Schem XV.

[51] Birch, *The History of the Royal Society*, Vol. I, p. 422.

[52] The anonymous treatise was published as a supplement to the Amsterdam edition of Willis's *Cerebri anatome* in 1664, 1666, 1667 and 1676, and subsequently in Willis's *Opera omnia* of 1676 and 1680.(See Raymond Hierons and Alfred Meyer, "Willis's Place in the History of Muscle Physiology," *Proceedings of the Royal Society of Medicine*, 57 (1964): Footnote 1, p. 658.

underlying the hypothesis.

   •The hypothesis stated and explained.

   •A demonstration of its mechanical application.

   •A discussion and restatement of the hypothesis in the light of additional operative factors.

A summary such as this credits Croone with a logical and analytical approach that is far from obvious even after several readings of his work: nevertheless, the basic schema is there, and there is evidence of a conscious attempt to fulfill the objectives of the Royal Society as expressed by Sprat, namely "bringing all things as near the Mathematical Plainness, as they can: and preferring the language of Artizans, Countrymen and Merchants, before that of Wits, or Scholars."[53]

Earlier Opinions as to the Cause of Muscle Movement

Croone notes two earlier opinions on the cause of muscular contraction. The first contention is that animal spirit is transmitted from brain to muscle, via the nerves, on command. A variant of this view is that the muscles are continuously supplied with animal spirits that also carry something akin to a motor faculty, thus rendering the muscle compliant to the command of the will. Croone would seem to be referring to one of Galen's several concepts; namely, that of the transmission of a quality like light or heat along the nerve in order to effect contraction and which Hall describes as "a 'flow of potency' through the resident pneuma [spirit]."[54] Croone disagrees with the more recent proponents of the animal spirit inflation theory, namely Descartes and Regius, on the grounds of incompatibility with anatomical evidence. There are no valves in nerves comprised of thread-like filaments, the size of the nerve is not commensurate with rapidly filling the muscle, and there are no inflatable cavities, as described, in the muscle.

The second opinion supposes that all muscles have an inherent tendency to contract— this being prevented by their arrangement around the joints such that oppositely directed pulls are balanced. Active contraction, and hence movement, occurs when one muscle receives some

---

[53] Thomas Sprat, *History of the Royal Society*, (London: 1667), edited by Jackson Cope & Harold Jones, (St Louis: Routledge, Kegan & Paul, 1959), p. 113.

[54] Thomas S. Hall, *Ideas of Life and Matter*, Vol. I, p. 162.

power or virtue [virtus] from the nerve, that enables it to overcome the strength of its antagonist. Croone attributes this opinion to Scarburgh, one of his contemporaries, but it appears to be in accord with Galen's *De motu musculorum* where that which stems from the nerve and gives a muscle the additional strength to overcome its antagonist is undefined, and is thus open to a number of interpretations including spirit flow. Croone adds that this opinion is confirmed by the muscle sectioning experiments first described by Galen and recounted by Fabricius, which demonstrated that muscle has the property of natural or inherent contractility.[55] Croone also makes the astute observation that, if you bend someone's arm against his will and he is resisting you as strongly as he can, the upper muscles swell in the same way as if the arm is being bent voluntarily (see Figure 3); surely, Croone argues, this cannot result from an animal spirit flow from the brain only at the command of the will. This apparently valid reasoning evidences the lack of existing knowledge concerning coordinated muscle activity. For Croone the elbow flexors are exerting their natural tendency to contract when shortened by an external force: in present day terms they are voluntarily contracting to stabilize the shoulder and elbow joints, thus enabling the elbow extensors to contract more effectively against the external force.

In general, Croone agrees with Galen that natural or inherent contractility plays no small part in effecting movement, but that it must be supplemented by some other force or power to enable a muscle to overcome the natural contractility of its antagonist and so move a part of the body. It is Croone's expressed intention not only to explain the nature of this super-added power, but to demonstrate that it complies with the laws of mechanics.

Presuppositions of the Hypothesis

i) Material Cause

Croone is anxious to avoid entangling himself in the question of how the mind or soul [anima] acts in relation to producing movement. The immediate cause of movement is material, as evidenced by convulsions or

---

[55] Galen demonstrated that cutting through a muscle from either a recently dead or living animal, resulted in the two segments retracting. Galen, *De motu musculorum*, p. 8; Fabricius, *De musculi actione*, p. 88.

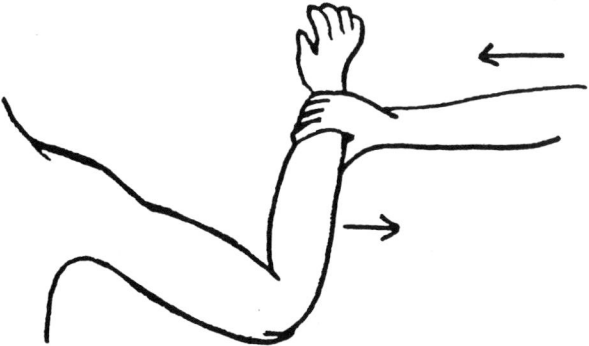

Bending someone's arm while they resist
the movement as strongly as possible

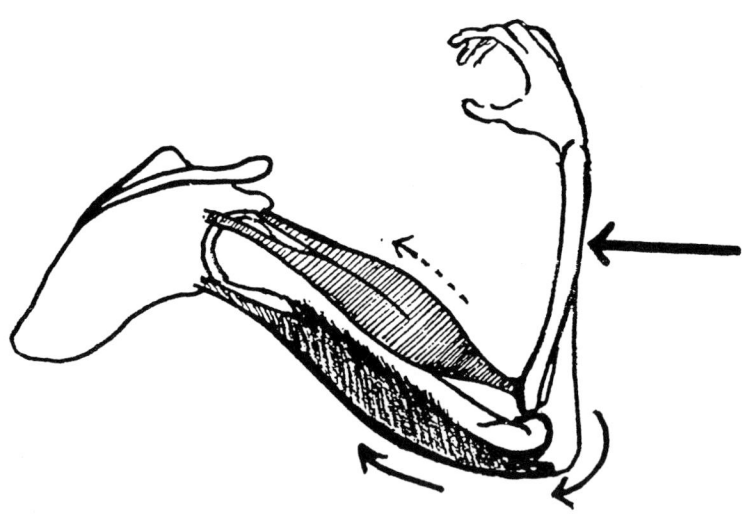

Flexor muscles swell up in the same was as if
voluntarily contracting to flex the forearm

Figure 3. To illustrate Croone's observation that "if you bend someone's arm against his will" and he is resisting you, the upper (flexor) muscles swell in the same way as if contracting voluntarily.

disordered movements that must occur independently of the mind. Only if it is claimed that the mind itself is affected by disease could it be thought to be responsible for potentially damaging movements.[56]

ii) Nerve Function

Croone reiterates the established facts that severing or ligating the nerve leads to loss of sensation or movement, or both. The site of the lesion adds further proof of the need for a continuous link between brain and muscle, since muscles receiving their nerve supply proximal to the injury are unaffected, whether the damage is to a particular nerve or to the spinal medulla that gives rise to the peripheral nerves. Croone further notes that:

a) the size of the nerve is directly proportional to the bulk, strength and frequency of activity of the skeletal muscle that it supplies.[57]

b) Nerves usually insert into the upper part of muscles and spread out downwards for, as Galen says, they contract only towards their head or proximal attachments.

c) Some very long muscles have a double nerve supply, with a branch entering the lower part of the muscle to supply that part which the upper nerve does not reach.

It is with this latter point that it is clear that Croone is closely following Fabricius, as reference to the text of *De musculi utilitatibus* confirms, but he has separated anatomical observation from conjectures as to the magnetic or luminescent nature of the nerve "force" that

---

[56] Convulsive movements, contrary to the will, contributed to Descartes' conclusion that "all the movements that we do not find by experiment to depend on thought, ought not to be attributed to the soul but merely to the disposition of the organs." (Descartes, "La description du corps humain . . . ," *Oeuvres de Descartes*, Adam and Tannery (Eds), (Paris: Léopold Cerf, Vol. X1, 1909), p. 225). Harvey also believed that "the muscles or motor organs, when in spasm and convulsions proceeding from some disturbing cause, are not moved otherwise than a cock or hen whose head has been cut off . . . such movements are always altogether confused and disordered, because the dominion of the brain has been removed." (Harvey, *Exercitationes anatomicae de generatione animalium*, as translated by Gweneth Whitteridge as *Disputations Touching the Generation of Animals*, (Oxford: Blackwell Scientific Publications, 1981), p. 18.

[57] A similar statement is found in Galen, *De usu partium*, as translated by Margaret Tallmadge May, *On the Usefulness of the Parts of the Body*, (Ithaca, N.Y.: Cornell University Press, 1968), p. 264.

Fabricius explores in some detail.[58]

iii) Nerve Structure

It is not easy to follow Croone's naked eye description of a nerve, but essentially he claims that it comprises innumerable tiny threads, surrounded by a moist medullar substance and enclosed by a double membrane. Within the muscle, the nerve divides into smaller and smaller branches that finally disperse into membranous folds and thus vanish away.

iv) The Spirits

In keeping with both Galenic and Cartesian systems, Croone believes that the spirits derive from the arterial blood. The repeated circulation of the blood mixed with chyle is responsible for removing the coarser and terrestrial particles, thus effecting a change: using chemical terminology, they pass from a state of fixation to a state of volatility. Croone does not explain how this volatility is achieved, but I assume that he is following the account given by Charleton in which the vital blood remaining in the heart after each contraction or systole "doth heat and kindle" the blood entering during relaxation or diastole: after several passages through the heart the spirits achieve the highest degree of volatility.[59] While the particulate nature of the blood is in accord with Descartes' opinion, Croone's description of the mercurial liquor, impregnated with salt and volatile sulphur, which is absorbed from the arteries into the medulla of the brain, takes origin from both Paracelsian chemistry and the more recent chemical analogies of Thomas Willis.

In the opinion of Charles Webster, "It is manifest that Paracelsus became a dominant influence in English medicine after 1640,"[60] and Debus confirms that "in the decade of the 1650s more Paracelsian and mystical

---

[58] Fabricius, *De musculi utilitatibus* (included in *De musculi artificio*), pp. 114-118.

[59] Charleton, *Natural History . . .* , p. 63.

[60] Charles Webster, "Alchemical and Paracelsian Medicine," Chapter 9 in Charles Webster (Ed), *Health, Medicine and Mortality in the Sixteenth Century*, (Cambridge; University Press, 1979), p. 317.

chemical works were translated than in the entire century before 1650."[61]
Le Fèvre (Nicaise Le Fèbvre), the Frenchman who was appointed
professor of chemistry to Charles II in 1660, and elected to the Royal
Society, may also have influenced Croone; in *A Compleat Body of
Chemistry*, 1660, Le Fèvre defined chemistry as

> nothing else but the Art and Knowledge of nature itself; that it
> is by her means we examine the Principles, out of which
> natural bodies do consist and are compounded, and by her are
> discovered unto us the causes and sources of their generations
> and corruptions, and of all the changes and alterations to which
> they are liable. . . .[62]

Paracelsus (1493-1541) had declared that all matter consisted of three
substances, the "tria prima" of mercury, sulphur and salt: specific
properties were ascribed to each "primal substance," sulphur being
combustible, mercury contributing liquidity, fusibility and volatility, and
salt being non-volatile and incombustible. The development of
Paracelsus's ideas from the Greek-Arabic theory of the constitution of
metals and other matter from mercury and sulphur, and the linking with
the medieval concept of body (salt), spirit (mercury) and soul (sulphur) is
succinctly summarized by Stillman.[63] In relation to the development of
chemistry in the sixteenth century Stillman contends that "the appeal of
the Paracelsian 'tria prima' to the chemists of the period lay in its more
comprehensible relation to experimental observation,"[64] and doubtless this
attitude prevailed in the following century.

In describing the nervous liquor as "mercurial" and "impregnated
with salt and volatile sulphur" Croone seems to be applying Paracelsian

---

[61] Allen G.Debus, *The Chemical Philosophy, Paracelsian Science and Medicine in the
Sixteenth and Seventeenth Centuries*, Vol. II, (New York: Science History Publications,
1977), p. 386.

[62] Le Fèvre, "A Compleat Body of Chemistry," (London, 1670; 1ˢᵗ French edition, 1660),
p. 1, as quoted by Debus, op. cit., p. 451.

[63] John Maxson Stillman, *The Story of Alchemy and Early Chemistry*, (New York: Dover,
1960), pp. 320-321. This is a reprint of his 1924 work entitled *The Story of Early Chemistry*.
More detailed accounts can be found in Walter Pagel, *Paracelsus: An Introduction to
Philosophical Medicine in the Era of the Renaissance*, (Basel: Karger, 1958), pp. 100-104; also
J. R. Partington, *A History of Chemistry*, vol. 2, (London: Macmillan, 1961), pp. 139-143.

[64] Stillman, *The Story of Alchemy and Early Chemistry*, p. 377.

concepts to Glisson's "succus nutritius" in order to account for its predicted subtlety, ability to flow slowly and gently along the nerves and potential to interact with arterial blood and thus inflate the muscle. With respect to the distillation of arterial blood, Croone may simply be following Glisson, using chemical terminology rather loosely, or he may be referring to the analogy developed by Thomas Willis in *De fermentatione*, 1659, where Willis says, "It seems to me that the Brain with the Scull over it, and the appending Nerves, represent the little Head or Glassie Alembic, with a Sponge laid upon it," the rarefied arterial blood distilling through the sponge or brain into the nerves, just as spirits of wine are rectified by boiling and condensation.[65] Once it reaches the nerves, the nervous liquor is diffused in all directions and eventually reaches the veins by which it is returned to the heart.

All parts of a living body are distended by spirituous liquors, the precise composition varying according to the nature of the part and the particular ferment it contains. Thus, Croone explains, within the whole muscle there is one type of spirit in the tendon and its fibers and another in the flesh of the muscle, neither of which is identical with that which flows in through the nerves. I am uncertain whether Croone is here following a particular physiological system, or himself adapting concepts put forward by Paracelsus and adopted by Jan Baptista Van Helmont (1577-1644), who maintained that every organ has its own archeus insitus or resident spirit.[66] Croone was certainly aware of Harvey's denial of any spirits apart from the blood since dissection failed to reveal them.[67] Nevertheless, Croone reasons that he is following Harvey's example in denying the existence of any "animal spirits" apart from the nervous fluid;

---

[65] Thomas Willis, "De fermentatione," 1659, in *The remaining Medical Works . . .*, p. 14.

[66] Debus, *The Chemical Philosophy . . .*, p. 360. Davis, *Circulation Physiology and Medical Chemistry in England*, p. 119 makes the point that Willis also believed that each organ had a unique ferment, ferment being any humor in which the particles of salt, sulphur, or spirit were exalted.

[67] "Even if the blood that flows in the arteries swells with a greater store of spirits, yet it is to be reckoned that these spirits are inseparable from the blood . . . and that blood and spirit make one body. . . . " Harvey, *Exercitatio anatomica de motu cordis et sanguinis in animalibus*, as translated by Gweneth Whitteridge in *An anatomical Disquistion Concerning the Movement of the Heart and Blood in Living Creatures*, Oxford: Blackwell Scientific Publications, 1976), pp. 13-14. For further discussion of Harvey's ideas on spirits, see Gweneth Whitteridge, *William Harvey and the Circulation of the Blood*, (London: Macdonald, 1971), especially pp. 223-226.

that visible blood is not analogous to invisible nervous fluid is glossed over by reiterating that the nerves are not hollow, which would be required for the transportation of large volumes of vaporous spirit from brain to muscle. They comprise instead a pithy substance appropriate for the transport of a slowly percolating fluid. Further evidence for the existence of spirituous liquors is adduced from the fact that an excess of spirits from drinking wine tends to quicken body movements, whereas a deficiency of spirits from disease or violent exercise causes fatigue. Croone does not proceed, however, to explain the mechanism of an alcoholic stupor.

v) Sensation and Nerve Fluid

Although a discussion of sensation is not strictly relevant to his topic, Croone believes that if the turgid, liquor–filled nerve membrane can be seen to transmit sensation, then this supports the proposition that the liquor-filled brain and nerve medulla have a motor function. Croone argues, firstly, that the medulla of the brain has no sensory function as evidenced by head wounds where brain substance has been painlessly removed. Secondly, Bartholin's account of an ox whose brain appeared to be turned to stone afforded proof that as long as passages exist for the animal spirit to travel from the cerebral arteries to the medulla of the nerves, then movement can occur.[68] Thirdly, in wounds to nerves, touching the medulla is painless, whereas the covering membrane is exquisitely sensitive to touch. Croone adds that anatomists have rightly asserted that all the sensitive membranes of the body, including those covering muscles, originate from the covering membrane of the brain.[69]

---

[68] Charleton was later to refer to this account of Bartholin and, in a rare display of humor and scepticism, observe that in his "desire to prop up the antique opinion of the Animal spirits . . . he was first so ingenious to suspect, and after so lucky as to find certain holes. . . . A wonderful providence of Nature this, to continue both the poor Animal in motion, and the doctrine of the spirits in reputation." Charleton, *Enquiries into Human Nature in VI Anatomic Prelectiones*, (London: 1680), p. 517.

[69] Harvey cites Fernel and Piccolomini as holding this opinion, namely that "the nerves convey the motive faculty in their medulla and the sensitive faculty in their coats that derive from the meninges (see *Prelectiones anatomie universalis*, edited and translated by Gweneth Whitteridge in *The Anatomical Lectures of William Harvey*, (Edinburgh: Livingstone, 1964), p. 239. Whitteridge's footnote, p. 238, adds that Fernel's argument was quite new and based on the finding that "delirium and lethargy, which are affections of the cerebrum, are not accompanied by pain, whereas the accumulation of humors in the meninges is exceedingly painful" thus proving that the meninges are sensitive to touch;

Accepting that these membranes are kept in a state of tension by the continuous perfusion of a characteristic spirituous liquor, then it is not difficult to accept that they are capable of vibrating like bells or very pure glass if struck. The resultant vibration is transmitted to the brain. Croone offers no comment on how vibrations travel through nerve plexuses where inter-connections are such that distortion would result. This model of sensation reaching the brain via the vibrating nerve membrane, and movement being effected by fluid ejection from a liquor–filled medulla could explain the clinical phenomena of loss of movement with intact sensation, and loss of sensation with intact movement. Where Harvey had noted these sensori-motor disturbances without offering any particular explanation,[70] Croone develops the vibration and fluid ejection model. The nerve membrane, impaired by injury or excess moisture, is likened to a cracked bell or glass where vibration is disturbed or lost. In sleep, Croone suggests that the peripheral circulation of the nervous liquor is reduced in the same way as is the circulation of the blood; hence, the limbs exhibit a greatly reduced capacity to move and respond to touch and the eyes, if opened, cannot see because of the greater laxity of the membranes.[71]

vi) Muscle Structure

Croone envisages the muscle as a structure of tendinous fibers, separated in the center of the muscle and woven together at the ends to form strong tendons. The intervals between the fibers are filled with flesh. The muscle is enveloped by membranes that contribute to the ease with which one muscle glides upon another. Nerves, arteries, veins and lymph ducts are also seen to be essential components.

Piccolomini held similar but not identical views. Croone refers also to the membrane that is spread out over all the muscles of the entire body, found in animals and believed to exist in man "below the fat"; according to Bauhin, "it is possessed of exquisite sensitivity" such that rigor results from it being pinched by some sharp humor. See Harvey, *Prelectiones . . .* , p. 61 and Footnote 4, p. 60.

[70] Harvey, *De motu locali animalium*, p. 103; p. 109.

[71] Galen, also, had considered whether muscles were inactive during sleep, and concluded that some activity was confirmed by one's ability to sleep on either side (not rolling as a corpse would), sleep with an object clasped in the hand, and with the mouth closed. See *De motu musculorum*, p. 18.

Croone lays particular emphasis on the fibers of the muscle, noting that:

a) they constitute the greater part of the muscle and are very strong. The gracilis muscle, dissected out from the inner side of the human thigh, could support a weight of 80 lb tied to its tendon without tearing.

b) their arrangement varies from muscle to muscle, according to functional requirements. The basic differences of parallel, oblique and spiral fibers or fasciculi are recognized, Croone being of the opinion that a more detailed study of this fiber arrangement could contribute to understanding the cause of contraction. There is no hint as to the nature of this contribution, and it may be that Croone was acknowledging some acquaintance with the work of Nicolaus Steno.[72]

c) The intervals between the fibers vary; as confirmed by Vesalius, the fibers are fleshier and more widely separated in that part of the muscle that swells on contraction, namely the muscle belly.[73]

vii) Spinal Medulla and Brain

Soaking the spinal medulla in water reveals its fibrous structure, from which Croone concludes that it functions in the same way as the nerves. The brain, also, is probably fibrous.

The Hypothesis Stated and Explained

It is indicative of a conscious methodology that Croone clearly states his intention to propose an hypothesis that can be rejected or developed, depending on whether it conforms with future experimental evidence. It is possible that this statement reflects Boyle's influence in view of his criteria for both a good and an excellent hypothesis, and his contention

---

[72] Croone's comment predates Steno's published work on muscles, but it is conceivable that Steno's interest in muscle structure had been indirectly communicated.

[73] While Croone defers to the authority of Vesalius to support this feature of muscle structure, essential to his hypothesis, he remains silent about Vesalius' opinion as to the flesh as "the principal organ of motion and in no wise a couch and support for the fibers." (Vesalius, *De humani corporis fabrica libri septem*, Brussels, 1964, p. 222. This is a facsimile reprint of the Basle 1553 edition, first published 1543).

that in physiology [natural philosophy]

> it is sometimes conducive to the discovery of truth, to permit
> the understanding to make an hypothesis, in order to the
> explication of this or that difficulty, that by examining how far
> the phenomena are, or are not, capable of being solved by that
> hypothesis the understanding may, even by its own errors, be
> instructed.[74]

Croone's conception of a muscle with its nerve and blood supply is
shown in his Figure 1 (p. 58). Croone reminds his reader of previous
assumptions, namely that the fibers and the flesh-filled spaces between
them are nourished and distended by spirituous fluids, specific to each and
evenly distributed: an impulse moves the fluid towards either end of the
muscle, either in a straight line or following the lines of the fibers, which
may be curved or spiral. Croone quotes from Fallopius to support the
orderly arrangement of the fibers and the consequent orderly disposition
of the spirits.[75]

Before proceeding to explain how the muscle contracts, Croone
decides, as did Descartes, to regard the body as sort of machine or
automaton, with the mind as a spectator of events. Convulsive movements and the movements of newborn infants are cited as evidence that
muscles can be moved without the influence of the mind. Coordinated
movements such as walking must be learned by practice and imitation and
are thus executed in a different way compared to disordered movements,
but Croone does not explain how the effective mechanism differs. Perhaps
the most interesting point about Croone's discussion is not his recognition
and shelving of the mind-body problem, but the difficulty he subsequently
experiences in avoiding speculation as to how the mind operated the
machinery.

The nature of the impulse initiating muscle contraction causes
Croone some concern. That it is within the power of the mind to move
only the nerve fibrils leading to a specific muscle is as believable as that it

---

[74] Boyle, *Complete Works*, 1772, Vol. I, pp. 302-303, as quoted by Laurens Lauden, "The
Clock Metaphor and Probabilism: the Impact of Descartes on English Methodological
Thought," *Annals of Science*, 2(2) (1966): p. 94.

[75] Gabriele Fallopio (1523-1562) was an anatomist and contemporary of Vesalius; his most
important work, *Observationes anatomicae*, from which Croone quotes, was published in
1561.

can transmit spirits into the nerve branch leading to that same muscle. Croone considers that if both sensation and ideas can be accounted for in terms of movement of the particles or fibers of the brain, then it can be granted that the impulse to move a leg or an arm is a similar movement that is transmitted to the appropriate motive fibers of the brain. Croone follows Descartes in his opinion that ideas and memory are material; as a physician Croone argues that unless they are material, they could not be confused or obliterated by disease. His discussion of this is far less complex, however, than that of Descartes who envisaged ideas as patterns traced or imprinted in the spirits on the surface of the pineal gland, and memory as these spirit-imprinted ideas traced onto the inner surface of the brain by a mechanical folding of the brain filaments and an enlarging of the spaces between the filaments. With memory, repetition reinforced this pathway that served to reform ideas.[76] The point Croone aims to establish is that if ideas can be spatially localized and linked within the brain, then the desire or impulse to move a part of the body can be conveyed to that place where the fibers connected to the relevant muscles end: movement of these brain fibrils results in the required movement. The link between idea of movement, brain area and execution of the desired movement has been established by learning to move and then repeating the movement.

To strengthen his hypothesis, Croone refers to sneezing and the constant swallowing that ensues when the uvula is enlarged and touches the esophageal membrane. Both these actions illustrate that there is an established and specific pathway between sensitive membrane, brain, and muscles—such that movement conveyed to the brain by way of the nerve membrane excites specific fibrils in the brain, which then move the muscles involved in sneezing or swallowing. Unless one accepts that acts such as sneezing and swallowing are carried out completely mechanically, then the alternative, according to Croone, is to assert that the mind spontaneously thrusts the animal spirits into the appropriate nerves and muscles. Since these acts are quite involuntary, a further question arises about the kind of skill by which the mind, perceiving the movements of the disturbed membranes, learns to propel the spirits into the appropriate muscles. To answer that the mind acts out of necessity amounts to saying

---

[76] There is a difficulty in knowing the extent to which Croone's ideas derive from Descartes. Although Croone does not acknowledge Francis Glisson in this discussion, he seems to have been influenced by him, in particular by his comments on perception and movement causation as expressed, not very clearly, in his Anatomy Lectures (British Library Manuscript Collection, Sloane MS 3306, pp. 163-164 in particular).

that it occurs mechanically because of the anatomical structure. Croone's discussion is not easy to follow, nor am I sure of the originality of his train of thought. It is, however, obvious that a mechanistic explanation had particular appeal in the light of the confirmed fibrous sub-structure of the nervous system, coupled with involuntary activities such as sneezing and swallowing. The problem of what to do with the mind, and whether the motor impulse is a fluid flow or a vibration, are subsidiary to the underlying assumption that a material mechanism exists.

Accepting that the nerve impulse can be best described as a shaking or vibration of the nerve fibers, Croone supposes that this results in the release of innumerable droplets of liquor or animal spirit throughout the entire muscle. This liquor then mixes with the spirits of the blood—resulting in a continuous agitation of all the spirituous particles present in every part of the muscle, as when spirits of wine and human blood are mixed. There are a number of other instances, derived from chemistry, which illustrate the agitation that occurs when two different liquors are mixed, namely oil of vitriol [sulphuric acid] and common water, or butter of antimony [antimony chloride] and spirit of nitre [nitric acid]. Moreover, Croone adds, blood has a tendency to froth up and ferment as evidenced, first, by the movement of the heart: whether the blood is expanded by a specific ferment or by a vital flame perpetually rarefying the blood, the main point is that it has the potential to expand. Second, Croone cites instances where the whole mass of the blood is fermented by an excess of seminal matter as seen in adolescent boys and others where, presumably, skin eruptions provide evidence of the underlying blood condition. Highmore's explanation of hysteria, where blood polluted by corrupt semen ferments and swells in the lungs, is also adduced as supporting evidence.[77]

Returning to muscle, Croone tackles the supposed objection that such a small quantity of nerve fluid could cause such an extensive reaction. Firstly, the droplets emerge from innumerable nerve endings throughout the muscle so that the fermentation is immediate and widespread. Secondly, since effective purges and emetics have been prepared from

---

[77] Nathaniel Highmore (1613-1685) studied medicine at Oxford and graduated as a Doctor of Medicine in 1642. The source for his opinion on hysteria is probably his *Exercitationes duae . . . de passione hysterica et de affectione hypochondriaca*, published in 1660. There seems to be a link between Highmore's suffocation associated with hysteria and the breathlessness associated with aortic aneurysms and aortic valve incompetence that may occur in syphilis.

substances that have lost little of their original weight after numerous infusions, it follows that quantity is not necessarily a prerequisite for efficacy.[78]

Croone's next step, having reasoned analogically that fermentation or swelling could result from a mixture of nerve fluid and blood, is to remind his reader of the structure of the muscle where compartments, broader in the belly of the muscle than at the ends, contain the mixing fluids. Although Croone does not make mention here of his earlier experiments of filling a bladder with water, and his conclusion that the least force exerted laterally could overcome considerable vertical resistance, it would seem that he has this in mind when he proposes that the agitated liquors move in straight lines towards the muscle ends, from whence they are turned back to accumulate in the middle of the compartment. Where Regius had specifically supposed straight line motion of the animal spirits for effective valve operation, it is difficult to know quite why Croone thinks that this motion is necessary in his schema, given that the compartments are already designed for central expansion.

Because he believes that the expansive reaction detailed above would be insufficient by itself to move the muscle, Croone considers other possible ancillary mechanisms. Initially he argues that if the mind controls the animal spirits in such a way that it can direct them into any nerve, then it could act in the same way to thrust blood into the artery supplying the muscle. That the soul can influence blood flow is evident from emotional states such as anger, happiness, shame, and passion, in which the blood supply to various parts of the body is altered. Further it can be argued that this happens mechanically as an unwilled accompaniment to the material idea in the brain that gives rise to the emotion. Therefore,

---

[78] Croone refers specifically to the crocus of metals, crocus being the end product of direct or indirect oxidation. The *Lexicon Technicum*, 1704, refers to "Crocus Metallorum" or Liver of Antimony made by firing equal parts of powder of antimony and saltpetre; the shining part is the "crocus" or "liver" that must be separated from the dross, washed and kept for use, "Of this usually is made the Emetick Urine, or Vinum Benedictum, by infusing and Ounce of the Crocus powdered in a Quart of Wine for 24 hours." (John Harris, *Lexicon Technicum*, (London: 1704), unpaginated. John Ward began a notebook in 1667 which records his attendance at a chemistry course at Oxford; he "compounded a number of emetic mixtures, many containing antimony as their primary ingredient." A later entry records that Willis "used sulphur of antimonie as a gentle vomiter." (Robert G. Frank Jr., "The John Ward Diaries: Mirror of Seventeenth Century Science and Medicine," *Journal of the History of Medicine and Allied Sciences*, 29 (1974): pp. 169-170.

Croone contends, it is not absurd to think that the same idea that excites the will to move a muscle and mechanically guides the animal spirits into the appropriate nerve, also acts through the nerve to the auricles of the heart to propel a greater quantity of blood through the appropriate artery. This analogy does not withstand close scrutiny since the branching arterial network is not the same as the nervous network with its individual fibers, but Croone makes no attempt to explain precisely how the heart could direct more blood into a specific arterial branch. Before his reader has time to consider this latest development, Croone proceeds to describe yet another mechanical and most likely way of assisting muscle contraction. As the muscle begins to swell and the particles of the combined spirituous fluids move quickly away from each other, there is less resistance to arterial flow and more space for a greater quantity of blood to enter more rapidly.

Croone adduces three basic proofs to support his belief that the arterial blood has an essential role in contraction. Firstly, the arteries and veins are distributed to muscles in the same way as are the nerves. Spigelius is said to have affirmed that they enter the muscle at its beginning or its mid-point and never at the end except in very long muscles that may receive additional branches from vessels nearby. Within the muscle the branches reach every part, the veins in particular having only a single membrane wall as compared to a double-walled sheath when outside the muscle.[79] Croone's second point is that the swelling of a muscle is similar to the swelling of the penis; this stiffens as the result of the influx of a large quantity of blood that expands when mixed with seminal matter derived ultimately from the nervous liquor. Fabricius is cited as the authority for this opinion, but incorrectly since he specifically used the example of the penis to demonstrate that muscle is not similarly

---

[79] Adrian Spigelius (van der Spiegel) (1567-1625) was a disciple of Fabricius, who studied at Padua and taught anatomy there from 1618 to 1624.

While it is true that the branches of the principal artery and nerve enter the muscle at a fairly constant area known as the neuro-vascular hilus, Spigelius has perhaps oversimplified the situation or failed to note the subsidiary arteries that "are generally present and enter at the periphery or close to the ends of the muscle." (*Gray's Anatomy*, 35th edition edited by R. Warwick and P. Williams, (Edinburgh: Longman, 1973), p. 483). In regard to the veins, they have three coats, as do the arteries, but the middle muscular coat of the vein is very weak. The three coats would not be distinguishable by the naked eye in the venules formed from the capillaries and the smaller veins. It is assumed that the thicker outer coat of the vein leaving the muscle gave Spigelius the idea that it was single-walled within the muscle.

distended and hardened by the inflow of vital arterial spirits.[80] Thirdly, Croone adds the opinion of Fallopius that both flesh and fibers are essential components of any part that exhibits voluntary motion, the vigor of the movements being partly a result of the heat of the flesh derived from the blood that is present in larger quantities when movement occurs. None of the above arguments is particularly convincing or conclusive, but Croone continues to elaborate on the role of the arterial blood. As the muscle fills with blood, its hardness and tension increase not only because its vein cannot remove the increased volume, but because the thin venous walls are compressed and their patency reduced by distension of the surrounding tissues.

In summary, as a result of the expansive reaction between the nervous fluid and the blood, the internal structure of the muscle, and an increased blood supply, a muscle swells, hardens and shortens.

The Mechanical Application of the Hypothesis

It is essential to Croone's hypothesis that the supposed swelling can be shown to be both efficient and sufficient to account both for muscle contraction and for activities that involve lifting not only the limbs but heavy weights. Croone's geometrical representation of the brachialis muscle acting to raise the forearm and flex the elbow is designed to demonstrate that only a very small force, acting horizontally, is required to overcome not only the vertical weight of the forearm but also any additional weight that might be held in the hand. (See Croone's Figure 2, p. 58 and the accompanying proof, p. 117.)

Having proved the efficiency of his proposed method of muscle contraction, Croone proceeds to consider whether there are any other possible operative factors. Perhaps the nerve juice has a sharpness so that it acts like purgative and emetic drugs that irritate and constrict the fibers of the stomach and intestine. Also, the slightest touch of an irritant fluid (possibly acid), given to Croone by Boyle, caused a dissected-out human thigh muscle to contract rapidly. From these examples Croone concludes that if the nervous fluid possesses an irritant quality, then this could suffice to stimulate a muscle to contract. It could, however, be regarded only as an auxiliary mechanism; when a strong contraction is needed to overcome the antagonist muscle, the expansive reaction between nerve

---

[80] Fabricius, *De musculi actione*, p. 101.

fluid and blood together with an increased blood supply are, for Croone, the major causes. It is interesting that Croone sidesteps the issue of whether muscle always contracts by the same mechanism in response to differing stimuli, and concentrates on the nature of the cause of contraction in relation to the magnitude of the response.

Croone is not unaware of the mechanical role attributed to the muscle flesh, but he contends that sheet-like muscles that do not move joints have no need of flesh to supply additional leverage for the fibers, nor do muscles moving very light structures such as the skin or the lips need flesh to provide a mechanical advantage to their action. Flesh is essential to muscle primarily as a locus of Croone's expansive reaction.

Additional Operative Factors

In his summing up, Croone adds several factors and dimensions that unfortunately complicate his initial hypothesis. The property of spontaneous contractility reappears as a third operative factor, in addition to the nerve juice and the blood, and Croone observes that an increase in muscle swelling is accompanied by an increase in the ability of the fibers to contract themselves. Although Croone attempts here to integrate the observed behavior of muscle with his proposed contractile mechanism, his explanation of contraction in geometrical terms treats the muscle fibers as passive, non-elastic threads of constant length. Now it seems that their shortening is assisted by an innate property to contract in direct proportion to the internally applied stretching force. Croone sees no incongruity in maintaining two kinds of contraction— one innate and one superimposed to supplement the innate—since this best explains voluntary movement in terms of one muscle of a balanced pair overcoming its opposite number.

Another operative factor, briefly noted above, is the heat of the blood entering the muscle. Croone observes that this is significant insofar as a heated muscle contracts to about a quarter of its original length. No further information is given on how the muscle was heated in order to demonstrate this effect, and I suggest that this observation derives from Fabricius who noted that boiling a whole muscle reduces it to approximately a quarter of its former length.[81]

Finally, the quantity of moisture in the muscle, in the form of blood

---

[81] Fabricius, *De musculi actione*, p. 98.

and nervous fluid, assists contraction in the same way as wet ropes have been shown to raise great weights.

Croone has, it seems, considered every possible contingency and it is tempting to see this as a tactical ploy whereby he has protected his central and more dubious fermentation hypothesis with an array of other possible and plausible auxiliary mechanisms, each of which would have attracted some support. Croone contends that the greatest barrier to acceptance of his hypothesis is the observed speed of contraction and relaxation. His concluding remarks, therefore, stress that every part of the body is filled with spirituous and vital juices so that striking the taut nerve fibrils in the brain immediately releases droplets from the distal nerve endings in the muscle. The analogy of a syringe is added to emphasize that the smallest pressure or stimulus results in a rapid release of fluid. Effervescence occurs instantly—as when water is mixed with oil of vitriol (sulphuric acid), and blood flows through the artery into the muscle as quickly as water from an open tap. The fibers contract as soon as the muscle begins to swell and so everything happens instantaneously. Relaxation is effected by the sudden cessation of the effervescence, with the spirits dissipating almost instantly through the muscle membrane (thus explaining why violent exercise causes sweating), and the blood flowing rapidly out of the muscle as pressure on the veins decreases and relatively more blood circulates through the fibers of the antagonist muscle. Blushing, pallor from fear, and penile erection are adduced as examples to demonstrate the speed with which the blood flow can change.

Croone ends his treatise with the belief that if the so far nebulous spirits found only in the blood and nervous fluid are in fact liquids, then what has been said may be close to the truth. There are, however, problems associated with accepting muscle contraction as the result of inflation by two reacting fluids. What controls the quantity of fluids mixed, and hence the force of contraction, and exactly how does the effervescence cease? Croone does not acknowledge these problems, nor does he attempt to explain how graduated strength and speed of contraction could be effected by his proposed mechanism. At no time has Croone recognized eccentric movement or controlled relaxation as an observed facet of muscle activity,[82] hence the problem of regulating

---

[82] Harvey referred to "the measured separating of ends in lengthening" in his notes on different types of muscle action; his examples included sitting down and "laying something down with the hand." Harvey, *De motu locali animalium*, pp. 141-143.

precisely the subsidence of the chemical activity, which in turn would require controlled reduction of nervous fluid and blood, remained outside existing considerations.

DE RATIONE MOTUS MUSCULORUM (1667)

The second edition of Croone's work is prefaced by a letter from him in response to the publisher's request for a corrected and enlarged edition. The request is declined, partly on the grounds of ill-health, but more importantly because Croone has learned that Nicolaus Steno, a Danish anatomist and "good friend" whom he had met at Montpellier in the winter of 1665-1666 and with whom he had discussed the topic of muscle at some length, was about to publish a second treatise on muscles.[83] Croone adds that he would have liked to know the details of Steno's contribution to "this very obscure matter" before revising his earlier work. His letter thus confirms that the work is unchanged from that of 1664, not printed on Croone's initiative, and uninfluenced by Steno's *Elementorum myologiae specimen*, which appeared in the same year. While it is tempting to conclude that Steno would have told Croone about Swammerdam's isolated muscle experiments that negated a noticeable increase in volume following contraction,[84] the dates of these experiments have not been clearly established, and may not have commenced until after 1665.[85] Certainly Croone would have been aware of Steno's disinclination to offer an opinion about the cause of contraction,[86] but in the extant correspondence that followed their Montpellier meeting there is no evidence that muscle was a topic of discussion—beyond Steno's

---

[83] A translation of this letter, dated July 3rd, 1665, is given in Wilson, "William Croone's Theory . . . ," p. 177 and also in T. Kardel, "Steno on Muscles," *Transactions of the American Philosophical Society*, 84 (1994): Pt 1, p. 26. A brief account of the meeting is found in Wilson, p. 165, and in F.N.L. Poynter, "Nicolaus Steno and the Royal Society of London," *Analecta Medico-Historica*, 3 (1968): p. 273.

[84] Jan Swammerdam, "Experiments in the particular motions in the muscles of the Frog . . . ," *The Book of Nature*, London, 1758, pp. 122-125. This and other papers were not published until 1737 by Boerhaave.

[85] T. Kardel, "Willis and Steno on Muscles: Rediscovery of a 17th Century Biological Theory," *Journal of the History of the Neurosciences*, 5 (1996): p. 102.

[86] Steno had clearly expressed this view in *De musculis et glandulis observationem specimen*, Amsterdam, 1664, p. 21.

account of the dissection of a tortoise that exhibited movement in response to touch some twenty-four hours following decapitation.[87]

THE HYPOTHESIS MODIFIED

In 1681 Croone's modification of his original hypothesis was published in the *Philosophical Collections* under the title of "An Hypothesis of the Structure of a Muscle, and the Reason of its Contraction: read in the Surgeon's Theatre Anno 1674,1675."[88] Croone explains that Hooke, as editor of the *Philosophical Collections*, had invited this paper because he intended to give an account of Borelli's book in the next issue. As it happens, both Croone's paper and an English translation of the preface of Borelli's *De motu animalium* appeared in the same issue. Croone points out that he has seen only "a sheet or two, and two or three schemes for explicating Muscular Motion, the very same with those I made use of, and the same Experiment of the Bladder applied in another Scheme. . . ." Although ignorant of the details of Borelli's work, the apparent similarity has caused Croone to place a higher value on his earlier hypothesis and the changes he had since made to it, and to realise that "if ever I reprint and enlarge that Pamphlet, De Ratione Motus Musculorum, I would not be thought to have made use of what was another's for my own."[89]

Croone's 1681 paper presents a summary of the modifications made to his 1664 hypothesis, largely to answer the objection that if the contracting muscle "swell'd like a Bladder blown up . . . we did not see any such conspicuous swelling in the belly of the Muscle."[90] Croone gives no indication as to the source of this objection. Although Steno's *Elementorum myologiae specimen* (1667) offered a geometrical proof that a contracting muscle could swell without the influx of new material, this and other arguments adduced by Steno seem to have had no impact on

---

[87] Birch, *The History of the Royal Society*, Vol. II, p. 102. Reference to the tortoise is also found in Steno's *Elementorum myologiae*. . . .

[88] *Philosophical Collections*, 2 (1681): pp, 22-25.

[89] Ibid., p. 25.

[90] Ibid., p. 22.

Croone's thinking.[91] In the account of Steno's work that Croone gave to the Royal Society meeting of February 13th, 1667/8, his recorded interest was on Steno's experiments, particularly that of ligating the descending aorta of a dog, thus producing muscle paralysis of the hindquarters, that was reversed on removal of the ligature.[92] Although this experiment strongly supported the significant role of the blood in muscle contraction, thus strengthening Croone's hypothesis, it was not reported in this light, nor did Steno himself accord it more than a passing reference.[93]

In 1669 Richard Lower in his *Tractatus de corde* had emphasized that a contracting muscle did not visibly swell, but rather became smaller and harder.[94] Later that same year Jonathan Goddard's somewhat crude experiment of immersing a man's arm in a container and observing the water level on contraction of the arm muscles must have thrown some doubt on volume increase. The recorded discussion at the Royal Society where this experiment was performed, centered on the fluctuation of the water level caused by arterial pulsation rather than the implications of what Oldenburg noted as an overall decrease in muscle dimensions.[95]

Maintaining that a contracted muscle does swell, albeit barely perceptibly, Croone supposes that each fleshy muscle fiber consists of innumerable tiny interconnecting bladders, all inflatable by the effervescent mixture of blood and nervous fluid that is strained through the coats of each little globule or bladder into its cavity. Croone here

[91] Nicolaus Steno, *Elementorum myologiae specimen seu musculi descriptio geometrica*. Florence, 1667. See also T. Kardel, *Steno on Muscles*, which contains both the Latin text and English translation of Steno's *Elementorum myologiae* . . . together with pertinent comments in Kardel's introduction.

[92] Birch, *The History of the Royal Society*, Vol. II, p. 247. Poynter, op.cit, pp. 275-277 details the sequence of events that followed the unsuccessful attempts of Lower and King to repeat this experiment, culminating in its successful demonstration at the Royal Society meeting of July 16th.

[93] This experiment is briefly mentioned in the first of two smaller dissertations appended to Steno's major work, namely "Carnis carchariae dissectum caput," *Elementorum myologiae* . . . p. 86.

[94] Richard Lower, *Tractatus de corde: item de motu colore sanguinis et chyli in eum transitu*, London, 1669. Facsimile edition, with a translation by K. J. Franklin, in *Early science at Oxford*, Vol. IX, R. T. Gunther (Ed), (Oxford: University Press, 1932), p. 77.

[95] Goddard's experiment was performed at the Royal Society meeting of April 1st, 1669; a written account, which was registered, was read at the meeting of December 16th, 1669. Birch, *The History of the Royal Society*, vol.II, p. 412: Register Book (Copy), Vol. IV, p. 95.

implies that the nourishing juice of the arterial blood and the nervous fluid are both extravasated into the spaces between the carnous or fleshy fibers, and mix prior to entering the globules. Once within the globule, their "constant agitations, ebullition or effervescence, which with the natural heat that is partly the cause, and partly the constant assister of their motion" distends the vesicles and holds them distended, consistent with life.[96] Details of this process, obviously refined from the first hypothesis where the effervescence took place in the fleshy spaces between the fibers, are said to have been given in the lectures, but are not included in Croone's 1681 summary. The mechanism for additional distension so that the fiber shortens is said to be similar to that proposed in the first hypothesis, and I can only assume that Croone here intends this to be effected by a further addition of the effervescent reactants from both arteries and nerves.

The geometrical construction of similar triangles in Figure 4B represents the special case of a fiber comprising four globules, as in Figure 4A. Croone proposes that in this instance, each globule will be distended by one quarter of the total distension of a fiber comprised of one single globule AKE. If it is then supposed that there are four thousand globules in a fiber, then each globule will be distended by one four-thousandth part of the single globule fiber, but the same weight will be raised though the same distance as by the distension of a single globule (see legend to Figure 4B). While Croone's geometrical demonstration is correct, it can only apply, as before, to a rigid system, and takes no account of possible fiber elasticity or contractility. Significantly, the hypothesis does not deny that some swelling takes place, and is perhaps deceptive in minimizing this: while one four-thousandth part of the swelling of a single globule is infinitesimally small, the number of fibers in a muscle, and hence the swelling, could be quite considerable. Croone adds that his lectures dealt with the fact that contraction is accompanied by some swelling, together with an explanation as to "how every Muscle in Contraction grows harder, denser and more compact." Just how Croone reconciled the apparently contradictory concepts of swelling and increased density and compactness is not known, but it seems that he had every intention of retaining his supposed expansive reaction in the muscle fiber in the light of evidence that the volume of a contracting muscle does not appear to

---

[96] Croone, "An Hypothesis of the Structure of a Muscle . . . ," p. 23.

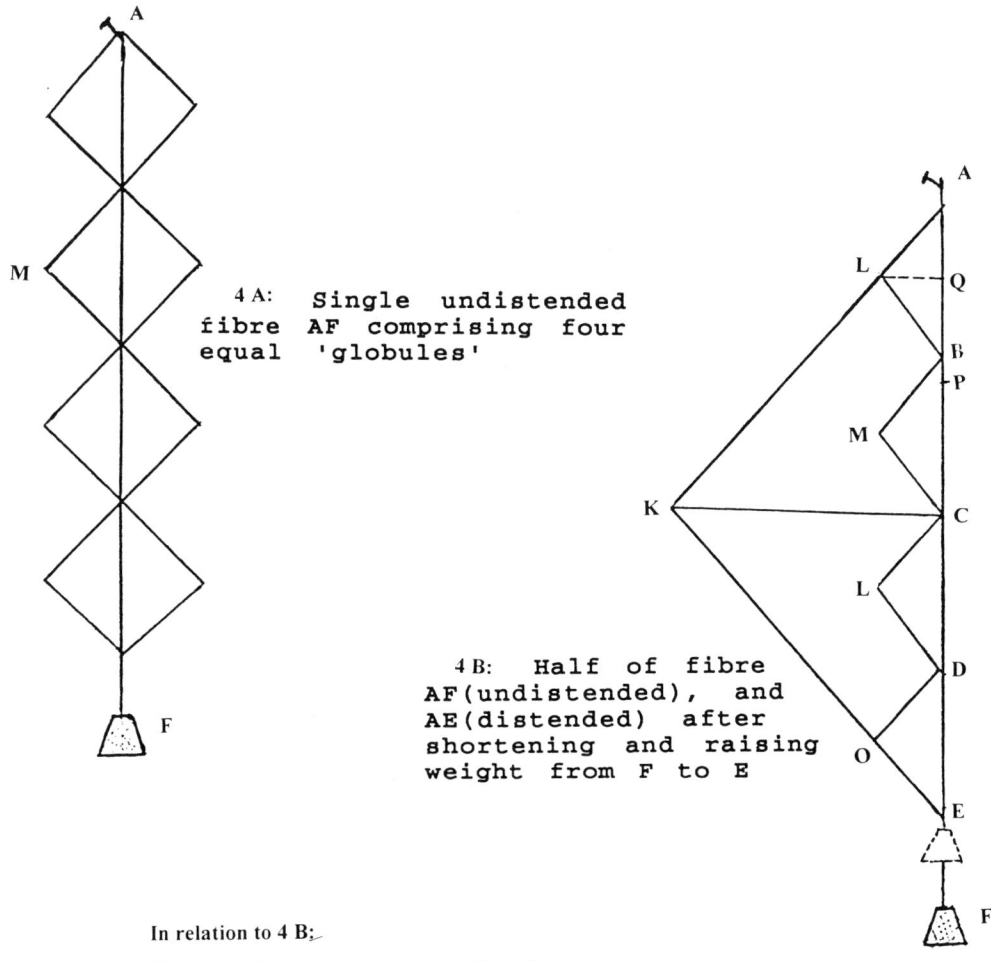

4 A: Single undistended fibre AF comprising four equal 'globules'

4 B: Half of fibre AF(undistended), and AE(distended) after shortening and raising weight from F to E

In relation to 4 B:

```
AEF = Carnous Fibre fixed at A.
ALB, BMC, CLD and DOE = Globules or Bladders, which
open into each other at points A,B,C and D.
Only 4 bladders are shown for the sake of simplicity:
in reality the number is 'infinite'.
Construction is such that △'s ALB = BMC = CLD = DOE.
ALB is similar to AKE, as are BMC, CLD and DOE.
AP = undistended lenght of AB.
AE = undistended length of AF.
BP and EF are in the relation  BP = 1/4EF, since AB =
1/4AE by construction.
Similarly QL = 1/4CK.
.. if AP shortens to AB and distance moved is BP =
1/4EF, it follows that BC,CD and DE will all have
shortened by a distance equivalent to BP = 1/4EF
.. Weight is raised by distance 4xBP = EF.
Also, since QL = 1/4CK, total distension is 1/4 that
which would occur if single fibre AKF were distended
to AKE.
.. if AKF consisted of 4000 bladders, a weight could
be raised through the same space EF with 1/4000th of
distension CK.
```

Figure 4. Geometrical demonstration of Croone's revised hypothesis of the structure of a muscle (redrawn from *Philosophical Collections*, 2, [1681]: Figure 3 and text, pp. 23-35).

increase to any perceptible degree.[97]

In concluding his paper, Croone notes that his hypothesis has gained strength not only from the apparent similarity with Borelli's schemes, but from the microscopic observations of Leeuwenhoek who has found the fleshy muscle fiber to comprise innumerable small vesicles or globules that "gives an appearance of reality to the said hypothesis."[98] At the Royal Society meeting of February 4th, 1674/5, when Edmund King read a paper on the supposed "tubes and liquor" construction of the animal body, including muscle, the recorded discussion reveals that Croone put forward his suggestion that the muscle fibers consist of chains of bladders "which being blown up by certain liquors shorten the said springs, and so contract the muscle." He was opposed by Hooke whose own observation was that the fleshy part of the muscle consists of an infinite number of exceedingly small round pipes, with movement resulting from filling and emptying these pipes.[99] Hooke, who later (1678) supposed that the muscle fiber resembled "a necklace of hollow glass beads," designed an experiment with a chain of small bladders fastened together to show how the fiber might be filled.[100] Thus, at the time of writing about his revised hypothesis, Croone could have been reasonably confident that microscopic observations had confirmed his suppositions concerning the fine structure of the muscle fiber. Interestingly, he makes no reference to Hooke's experiment and the ensuing discussions at the Royal Society, which considered issues such as the nature of the material that filled and expanded the bladders of the muscle fibers, and whether heat could effect expansion at the requisite speed. When Leeuwenhoek wrote to the Royal Society in February 1681/2, admitting that his earlier observations of muscle fibers were incorrect as the globules now appeared to be rimples [wrinkles],[101] Croone is not recorded as offering any comment in the discussions and

---

[97] This point has also been made by Troels Kardel in *Steno on Muscles*, pp. 26-27.

[98] Ibid., p. 25. Leeuwenhoek's communications to the Royal Society are to be found in the *Philosophical Transactions*, Vol. 9, No. 106, p. 122 (1674) and Vol. 12, No. 136, p. 899 (1677).

[99] Birch, *The History of the Royal Society*, Vol. III, Meeting of February 4th, 1674/5, pp. 179-180.

[100] Ibid., Meeting of April 18th, 1678, p. 401.

[101] *Philosophical Collections*, February 1681/2, pp. 152-160.

microscopy demonstrations that followed this communication.

Borelli gives few clues as to the source of his ideas on muscle contraction, but it is clear from some of his comments that he had read Croone and paid particular attention to his analysis of contraction in terms of horizontal forces within the muscle compartments,[102] and also to issues such as whether the blood supply to a contracting muscle could be increased by a simple act of the will,[103] as Croone had supposed. Nevertheless, Borelli's muscle fiber, comprised of a chain of rhombuses, does differ from that proposed by Croone in that the rhombuses do not physically interconnect, and are much larger than those envisaged by Croone.

In common with both Croone and Willis, Borelli proposes that a chemical reaction between blood and nerve fluid causes inflation of the rhombuses, although he fails to explain how this reaction accorded with his notion of the fast-moving wedge-shaped particles necessary to inflate the rhombuses with a percussion-like force.

Although the Royal Society discussed Borelli's work at some length,[104] there is no record of what Croone thought of his ideas. What we do know is that Croone continued to puzzle over muscle structure and function, and in a matter of months before his death in 1684, he sent the muscle of a man to Denis Papin to be placed in a receiver in order to determine the quantity of air produced "in vacuo." The experiment was interrupted when a crack developed in the receiver, and there is no documentation of the final result.[105]

---

[102] Joh. Alphonsi Borelli, *De motu animalium*, (Rome: Angelo Bernabo, 1680 pars prima and 1681 pars secunda), I, Propositions 95-99, pp. 181-186. This work has been translated by Paul Maquet, *On the Movement of Animals*, (Berlin: Springer Verlag, 1989).

[103] Ibid., II, Proposition 21, p. 50.

[104] Birch, *The History of the Royal Society*, Vol. III, Meetings of April 5th and July 26th, 1682, pp. 140 & 157.

[105] Birch, *The History of the Royal Society*, Vol. IV, p. 320. Some ten years earlier, Isaac Newton had suggested that Boyle should investigate the effect of compression or rarefaction on a muscle in his air pump, presumably with the aim of further elucidating Newton's ether hypothesis with respect to muscle contraction.

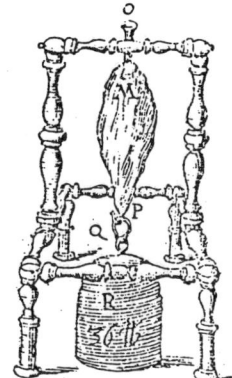

OM = wooden channel opening
     into bladder
M = leather valvule or flap
     opening inwards to prevent
     backflow of air
P = base of bladder
Q = hook suspending weight
R = weight (36 lb)

Figure 5 A: Device for raising a weight with a single
             bladder (Fig. 34 from *Collegium Experimentale*
             sive Curiosum p.188).

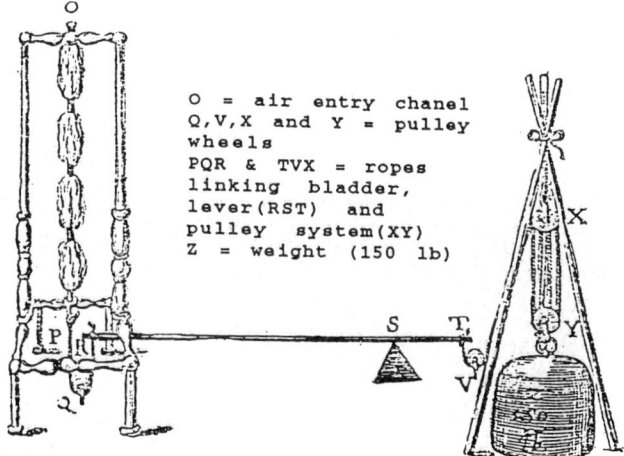

O = air entry chanel
Q,V,X and Y = pulley
wheels
PQR & TVX = ropes
linking bladder,
lever(RST) and
pulley system(XY)
Z = weight (150 lb)

Figure 5B: Device for raising a weight of 150 lb by inflating
            a series of four bladders (Fig.35 from *Collegium*
            Experimentale sive Curiosum p.189).

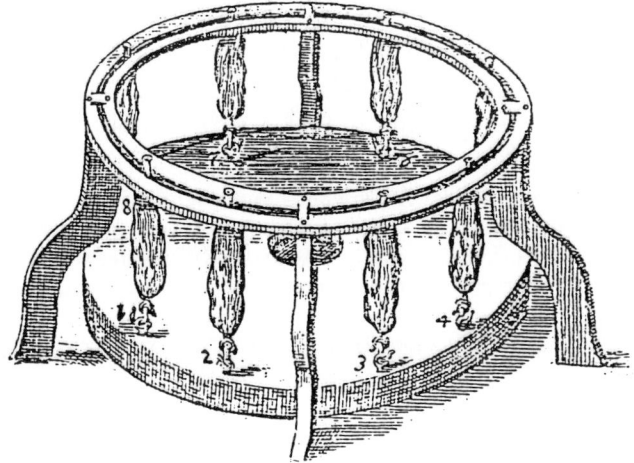

Figure 5C: Device for lifting a stone attached to the bladders as shown. This was
            to be operated by eight "robust youths" each blowing into one of the
            eight bladders at the same time (Fig. 36 from *Collegium Experimentale*
            sive Curiosum p.191).

Figure 5. Lifting devices utilizing inflated bladders from Joh. C. Sturm,
Collegium Experimentale sive Curiosum, Nuremberg, 1676-1685.

MECHANICAL DEVELOPMENTS
OF THE NEW FORCE OF INFLATION

While Croone's work seems to have had little impact in England,[106] it did not go unnoticed on the Continent.

One of the most intriguing developments of the Royal Society's bladder-inflation experiments and the inflation theory attributed to both Borelli in Italy and Croone in England is found in an account of the experiments undertaken at the Collegium Curiosum sive Experimentale in Nuremberg, and published in the second part of *Collegium Experimentale sive Curiosum,* 1685, edited by the founder of the college, Johann C.Sturm.[107] Figures 5A, 5B and 5C demonstrate how the "new force," which is "nothing else than the bladder of a pig or cow that can be shortened by breathing (blowing) of the mouth alone," can be used to lift a considerable weight. England is acknowledged as the source of the first experiments undertaken by the Collegium Curiosum, with some modifications such as the addition of a valve to the mouthpiece to prevent the backflow of air. In regard to the "the usefulness of this curiosity" it is "greatest in explaining the way in which muscles in bodies of animals are wont to move joints and bones and by mean of these to lift, carry, drag or otherwise move from their places huge weights."

Croone's geometrical solution is explained "in our own words," and the objection raised that "if a single muscular fiber, composed of four thousand little bladders, swells four thousand times less than if it were one continuous bladder" then a muscle containing four thousand fibers, èach of which has one four-thousandth part of the swelling "will have the same [swelling] as if it were a single bladder instead of a muscle composed of four thousand fibers." In considering this objection, it is pointed out that

---

[106] Kardel has considered Steno's influence on Croone's contemporaries, namely Lower, Willis and Mayow (T. Kardel, *Steno on Muscles,* pp. 28-33). There is no explicit mention of Croone's work by these investigators.

[107] Johann C. Sturm, *Collegium Experimentale sive Curios*um (Nuremberg: 1676-85. 2 vols in one, second part, 1685), Tentamen XI, pp. 187-204. Sturm was professor of mathematics and physics at the University of Altdorf for thirty-four years, and was considered the most skillful experimenter in Germany at the end of the seventeenth century. He formed the Collegium Curiosum in imitation of the Accademia del Cimento with the express purpose of doing wonderful experiments. See Martha Ornstein, *The Role of Scientific Societies in the Seventeenth Century,* (Chicago, Illinois: The University of Chicago Press, 3rd edition, 1938), pp. 175-177.

while a contracted muscle swells, it does not swell as much as the objection demands, nor is it reasonable to suppose that the number of fibers in a muscle equals the number of bladders in a fiber since the length of a muscle usually exceeds its thickness. Further, the softness and laxity of a relaxed muscle suggests that there is room for the bladders or globules of the fibers to expand without "the distensions of the fibers overflowing into the periphery of the muscle." Finally, it can be supposed that the fibers are so arranged, in parallel, that they interlock; "the bellies of the little bladders in one fiber correspond to the places where the little bladders meet in the adjacent fiber, and vice versa" thus reducing the overall expansion.[108]

In order "to make this whole thing still easier to be actually seen," there follows a description of the making of a model of a muscle fiber. Bovine intestine proved unsuccessful, and so a tube of close-weave linen was smeared with wax and constricted at intervals into little intercommunicating bladders. When inflated, the tube noticeably shortened with "small expansions regarding width." From this model the idea was developed of constructing a whole muscle of thin leather or waxed tafetta tubes enclosed within a common membrane, fastened at one end to form the tendon and gathered together at the upper end so that each tube or fiber could be evenly inflated by blowing into a single orifice. Indeed, a "whole machine rivaling the human body" could be constructed, the brain being replaced by air that could be siphoned through a non-return valve into a channel made of flexible material like leather to inflate specific muscles. Wires could be added to "delude its spectators with some semblance of perception" such that if touched, they would activate the mechanism distributing air to the muscles, and the model would respond "as if it were going to punch or kick the person doing the touching."[109] Thus, referring to Figure 6, pressure on the umbilicus causes the forearm of the statue to bend towards the chest. Details such as covering the artificial muscles with something resembling skin, attaching the muscles across joints without causing unnatural bulging, and allowing the air to escape gradually to simulate muscle relaxation are left to the "dexterity and talent" of those with the necessary leisure, skill, and patronage to perfect the proposed machine. Mention is made of the working model of

---

[108] Ibid., pp. 199-200. The brachialis muscle depicted in Figure 6 illustrates this interlocking.

[109] Ibid., pp. 200-202.

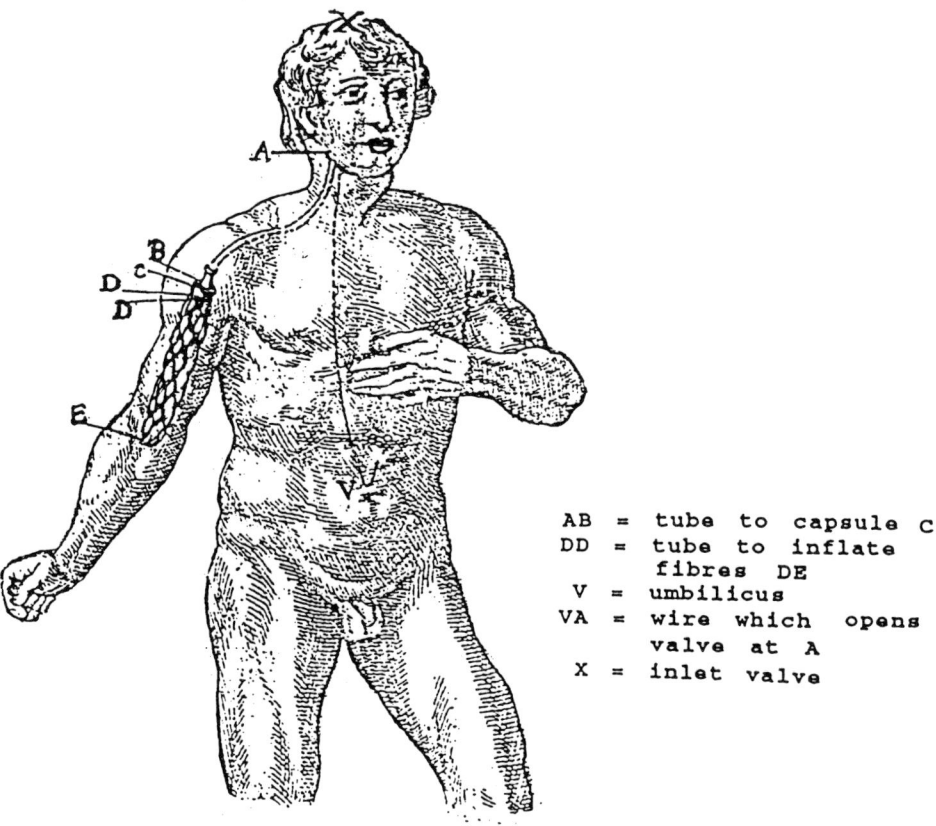

```
AB  =  tube  to  capsule  C
DD  =  tube  to  inflate
       fibres  DE
 V  =  umbilicus
VA  =  wire  which  opens
       valve  at  A
 X  =  inlet  valve
```

Figure 6. Model with "inflatable muscle" that bends the forearm when activated by pressing the umbilicus (V): this pulls on the wire VA which opens the valve at A, and releases air into the artificial muscle fibers, DE (Fig. 38 from *Collegium Experimentale sive Curiosum*, p. 203).

the circulation constructed by Reyselius[110] that is seen to have confirmed the possibility of building a mechanical contrivance to simulate muscle action.

Sturm's experiments and ideas utilizing the new force of inflation are, as the author states, curious rather than practical and provide an interesting response to criticism of the bladder inflation experiments as doing little "to direct vague empirical tendencies into productive experimental channels."[111]

IN CONCLUSION

In summary it can be said that Croone resolved the problems confronting the proponents of contraction by inflation insofar as his hypothesis could account for speed of action, minimal quantity of inflationary reactants, and the strength of contraction as a logical consequence of internal muscle structure. He also materialized the animal spirits by proposing that they were constituents of the blood and nervous fluid and therefore, to use Wilson's words, "susceptible to observation and reason."[112]

Because Croone's hypothesis is an eclectic amalgam of ideas, it supports a variety of descriptions from Galenic to Cartesian and from iatrophysical to iatrochemical. In essence, it is a pivotal work that attempts to take into account, albeit selectively, all the information then available on the structure and function of muscle and nerve, and to develop an explanation of muscle contraction in terms of what seemed to be appropriate chemical and mechanical concepts derived from a verifiable experimental basis. I have no doubt that part of Croone's work derived from discussions with colleagues arising from a long-standing belief that expanding bladders lifting weights provided a possible model to explain how a muscle fiber could shorten. In a relatively short time Croone's rudimentary conception of the vesicular microstructure of the muscle fiber was totally eclipsed by Borelli's detailed exposition, which in turn became the subject of further elaboration and analysis by the Bernoullis

---

[110] Reyselius designed a model of a mechanical circulation in 1677.

[111] Theodore Brown, *The Mechanical Philosophy* . . . , p. 66. The criticism is specifically directed at Wilkins's blown bladder experiments.

[112] L. G. Wilson, "William Croone's Theory of Muscular Contraction," p. 164.

(Johann[113] and Daniel), Keill, Pemberton and others. As Pemberton rightly observed, the vesicular hypothesis was well received largely because of "the extreme difficulty of conceiving any other means that might easily produce the effect under consideration."[114] Not until more refined mathematical analysis demonstrated that the degree of swelling and the forces developed were incompatible with reality,[115] and microscopy proved that there was no basis for "a series of vesicles" or "a chain of rhombs" in the muscle fiber,[116] was the vesicular hypothesis abandoned.

While Croone's work has not been entirely neglected by historians, it is his initial mechanism of contraction that has attracted most attention, often in the light of a revival of the ancient theory of inflation by animal spirits. This does less than justice to Croone who struggled with so many difficult aspects of muscle action, and endeavored to base his ideas on observations and experiments as opposed to philosophical imaginings. His work is not without loose ends and inconsistencies, but it does make interesting reading, more particularly since three hundred years were to pass before electron microscopy led to the development of the sliding filament theory and the provision of an acceptable mechanical and chemical explanation for muscle fiber contraction.

---

[113] Johann Bernoulli's "On the Mechanics of the Movement of the Muscles," translated by Paul Maquet, with an Introduction by Troels Kardel, is published in *Transactions of the American Philosophical Society*, 87 (1997): Pt 3.

[114] Henry Pemberton, in the Introduction to *Myotomia Reformata; or, an Anatomical Treatise on the Muscles of the Human Body*, the late William Cowper (London: 1724), p. xxiv.

[115] Ibid., pp. xxxiv ff.

[116] Albrecht von Haller, *First Lines of Physiology*, *The Sources of Science*, Vol. 32. (New York: Johnson Reprint Corporation, 1966, first published 1786), Proposition CCCCV.

REFERENCES

Birch, Thomas, *The History of the Royal Society*. New York: Johnson Reprint Corporation, 1968. A facsimile reprint of the first edition, London, 1756-7.

Borelli, Joh. Alphonsi, *De motu animalium*, Rome: Angelo Bernabo, (pars prima) 1680; (pars secunda) 1681.

[Boyle, Robert], *The Works of the Honourable Robert Boyle*. 5 vols. London, 1744.

Brown, Theodore M., "The Mechanical Philosophy, and the 'Animal Oeconomy': A Study in the Development of English Physiology in the Seventeenth and Early Eighteenth Century." Ph.D. thesis, Princeton University, 1968.

Bulwer, John, *Pathomyotomia or a Dissection of the Significative Muscles of the Affections of the Minde*. London, 1649.

Charleton, Walter, *Oeconomia animalis: novis in medicina hypothesibus superstructa et mechanice explicata*. London, 1659.

Charleton, Walter, *Natural History of Nutrition, of Life, and of Voluntary Motion*. London, 1659.

Charleton, Walter, *Enquiries into Human Nature in VI Anatomic Prelectiones*. London, 1680.

Crooke, Helkiah, *Microcosmographia, A Description of the Body of Man Together with the Controversies and Figures thereto belonging*. 2nd ed., London, 1631; 3rd ed., London, 1651.

Croone, William, "An Experimentall Account of the raising up of a weight hung at the bottome of an emptie Bladder," *Register Book (Original) of the Royal Society*. 1 (1661-1662): 109-112.

Croone, William, "An Hypothesis of the Structure of a Muscle, and the Reasons of its Contraction; read in the Surgeon's Theatre, Anno 1674, 1675," *Philosophical Collections*, Number 2, (1681): 22-25.

Davis, A. B., *Circulation Physiology and Medical Chemistry in England 1650-1680*. Lawrence, Kansas U.S.A: Coronado Press, 1973.

Debus, Allen G., *The Chemical Philosophy: Paracelsian Science and Medicine in the Sixteenth and Seventeenth Centuries*. 2 vols. New York: Science History Publications, 1977.

Descartes, René, *De homine figuris, et latinate donatus a Florentio Schuyl*. Lugduni Batavorum, 1662.

[Descartes, René], *Oeuvres de Descartes*. Charles Adam and Paul Tannery (Eds). Paris: Léopold Cerf, Vol. XI, 1909.

Dewhurst, Kenneth, *Thomas Willis' Oxford Lectures*. Oxford: Sandford Publications, 1980.

Fabricius ab Aquapendente, Hieronymus, *De musculi fabrica, De musculi actione* and *De musculi utilitatibus*. Vincenza, 1614.These three treatises are contained within the title *De musculi artificio*, published with other titles as part of a composite volume in Padua in 1625.

Frank, Robert G., Jr., "The John Ward Diaries: Mirror of Seventeenth Century Science and Medicine," *Journal of the History of Medicine*, 29(2) (1974): 147-179.

Frank, Robert G., Jr, *Harvey and the Oxford Physiologists: A Study in Scientific Ideas*. Berkeley: University of California Press, 1980.

Galen, *De motu musculorum*. Translated by Charles Mayo Goss as "On the Movement of Muscles by Galen of Pergamon," *American Journal of Anatomy*, 123 (1968): 1-25.

Galen, *De usu partium*. Translated by Margaret Tallmadge May as *On the Usefulness of the Parts of the Body*. 2 vols. Ithaca, N.Y.: Cornell University Press, 1968.

Galileo Galilei, *Dialogues Concerning Two New Sciences*. Translated by Henry Crew and Alfonso de Salvio. New York: Dover Publications, 1914.

Galileo Galilei, *On motion and Mechanics*. Translated by I. E. Drabkin and Stillman Drake. Madison: The University of Wisconsin Press, 1960.

Glisson, Francis, *Anatomy Lectures*. British Library Manuscript Collection, Sloane MS 3306. n.d.

Gray's Anatomy. 35th ed. Edited by R. Warwick and P. Williams. Edinburgh: Longman, 1973.

Hall, T. S., *Ideas of Life and Matter*. Vol. 1, *From Pre-Socratic Times to the Enlightenment*. Chicago: University of Chicago Press, 1969.

von Haller, Albrecht, *First Lines of Physiology*, The Sources of Science, Vol. 32. New York: Johnson Reprint Corporation, 1966. First published 1786.

Harris, John, *Lexicon Technicum, or a Universal Dictionary of Arts and Sciences*, Vol. 1. London, 1704.

Harvey, William, *Prelectiones anatomie universalis*. Edited and translated by Gweneth Whitteridge. Edinburgh: Livingstone, 1964.

Harvey, William, *De motu locali animalium 1627*. Edited and translated by Gweneth Whitteridge. Cambridge: University Press, 1959.

Harvey, William, *Exercitatio anatomica de motu cordis et sanguinis in animalibus*. Translated by Gweneth Whitteridge as *An Anatomical Disputation Concerning the Movement of the Heart and Blood in Living Creatures*. Oxford: Blackwell Scientific Publications, 1976.

Harvey, William, *Exercitationes anatomicae de generatione animalium*. Translated by Gweneth Whitteridge as *Disputations touching the Generation of Animals*. Oxford: Blackwell Scientific Publications, 1981.

[Harvey,William], *The works of William Harvey, M.D.* Translated by Robert Willis. New York: Johnson Reprint Corporation, 1965. A reproduction of the Sydenham Society, London, 1847 edition.

Hierons, Raymond and Meyer, Alfred, "Willis's Place in the History of Muscle Physiology," *Proceedings of the Royal Society of Medicine*, 57 (1964): 687-692.

Hooke, Robert, *Micrographia, 1665*. Reprinted in R. T. Gunther (Ed.), *Early Science at Oxford*, Vol. XIII. Oxford: University Press, 1938.

Hunter, Michael, *Science and Society in Restoration England*. Cambridge: University Press, 1981.

Kardel, Troels, "Steno on Muscles," *Transactions of the American Philosophical Society*, 84 (1994): Pt. 1.

Kardel, Troels, "Willis and Steno on Muscles: Rediscovery of a 17th Century Biological Theory," *Journal of the History of Neurosciences*, 5 (2) (1996): 100-107.

Lower, Richard, *Tractatus de corde: item de motu colore sanguinis et chyli in eum transitu*. London, 1669. Facsimile edition, with a translation by K. J. Franklin, in *Early Science at Oxford*, Vol. IX, R. T. Gunther (Ed.). Oxford: University Press, 1932.

Middleton, W.E.K., *The Experimenters: A Study of the Accademia del Cimento*. Baltimore: Johns Hopkins Press, 1971.

Munk, William, *The Roll of the Royal College of Physicians*. 2nd ed., London, 1878.

Nayler, Margaret A., "A Thorny Problem: Galen, Fabricius and Harvey on Muscles," unpublished M.A. thesis, University of Melbourne 1975.

Ornstein, Martha, *The Role of Scientific Societies in the Seventeenth Century*. 3rd ed., Chicago: University of Chicago Press, 1938.

Parsons, James, "The Croonian Lectures of Muscular Contraction," 1744 and 1745, *Philosophical Transactions*, Supplement to Vol. XLII.

Payne, L. M., Wilson, L. G. and Hartley, Sir H., "William Croone, F.R.S.(1633-1684)," *Notes and Records of the Royal Society of London*, 15 (1960): 211-18.

Pemberton, Henry, Introduction to William Cowper, *Myotomia reformata*. London: 1724.

Philosophical collections. Edited by Robert Hooke (1672-82) when publication of Philosophical Transactions suspended.

Poynter, F.N.L., "Nicolaus Steno and the Royal Society," *Analecta Medico-Historica*, 3 (1968): 273-280.

The Record of the Royal Society of London. 3rd edition. London: Oxford University Press, 1912.

Regius, Henricus, *Fundamenta physicis*. Amstelodami, 1646.

Robinson, Henry W., and Adams, Walter (Eds), *The Diary of Robert Hooke, M.A. M.D. F.R.S. 1672-1680*. London: Taylor & Francis, 1935.

Sprat, Thomas, *History of the Royal Society. London, 1667*. Edited by Jackson Cope and Harold Jones, St Louis: Routledge, Kegan & Paul, 1959.

Steno, Nicolaus, *Elementorum myologiae specimen seu musculi descriptio geometrica*. Florence, 1667.

Stillman, John Maxson, *The Story of Alchemy and Early Chemistry*. New York: Dover Publications, Inc., 1960. Reprint of 1924 work entitled "The Story of Early Chemistry."

Sturm, Johann C., *Collegium Experimentale sive Curiosum*. 2 vols in 1. 1676-1685.

Swammerdam, Jan, *The Book of Nature*. Translated from Dutch and Latin original by Thomas Floyd, Revised and improved by Notes from Réaumur and others, by John Hill, M.D. London, 1758.

Vesalius, Andreas, *De humani corporis fabrica libri septem*. Brussels, 1964. A facsimile reprint of the Basle, 1553 edition (first published 1543).

Wallis, John, "An account of the Experiment wherein a Weight is raised by the Blowing of a Bladder," *Register Book (Original) of the Royal society*. 2 (1662-63): 120-137.

Webster, Charles, *The Great Instauration. Science, Medicine and Reform, 1626-1660*. London: Duckworth, 1975.

Webster, Charles (Ed), *Health Science, Medicine and Mortality in the Sixteenth Century*. Cambridge: University Press, 1979.

Whitteridge, Gweneth, "Of the Local Movement of Animals: The Wilkins Lecture, 1979," *Notes and Records of the Royal Society of London*, 34 (2) (1980): 139-153.

[Willis, Thomas], *The Remaining Medical Works of that Famous and Renowned Physician Dr Thomas Willis...* Englished by S. P. Esq [Samuel Pordage]. London; T. Dring et al, 1681.

Wilson, L. G., "William Croone's Theory of Muscular Contraction," *Notes and Records of the Royal Society of London*, 16 (2) (1960): 158-178.

WILLIAM CROONE

———— ••◦◦€€€◦◦•• ————

ON THE REASON OF THE MOVEMENT OF THE MUSCLES

Translated from Latin by
Paul MAQUET, M.D.,
doct.h.c. Univ.Paris XII, doct. med. h.c. R.W.T.H. Aachen

Translation revised by
August ZIGGELAAR, S.J.,
dr. phil. Copenhagen

and edited by
Margaret NAYLER, Ph.D.,
The University of Melbourne

DE RATIONE MOTUS Musculorum.

Auctore ...

Ἐι πᾶσι τοῖς φυσικοῖς, ἐστί τι θαυμαϛόν.
Arift. de Part. Animal.

LONDINI,
Excudebat *J. Hayes* : Proftant Venales apud *S. Thomfon,* ad Infigne Epifcopi, in Cœmeterio Paulino, 1664.

On the reason

of the movement

of the muscles

"There is something admirable in all physical matters"

Arist. de Part. Animal.

LONDON

Publisher: *F. Hayes*. For sale by *S. Thomson,* at the sign of the Bishop, in St Paul's Cemetery, 1664.

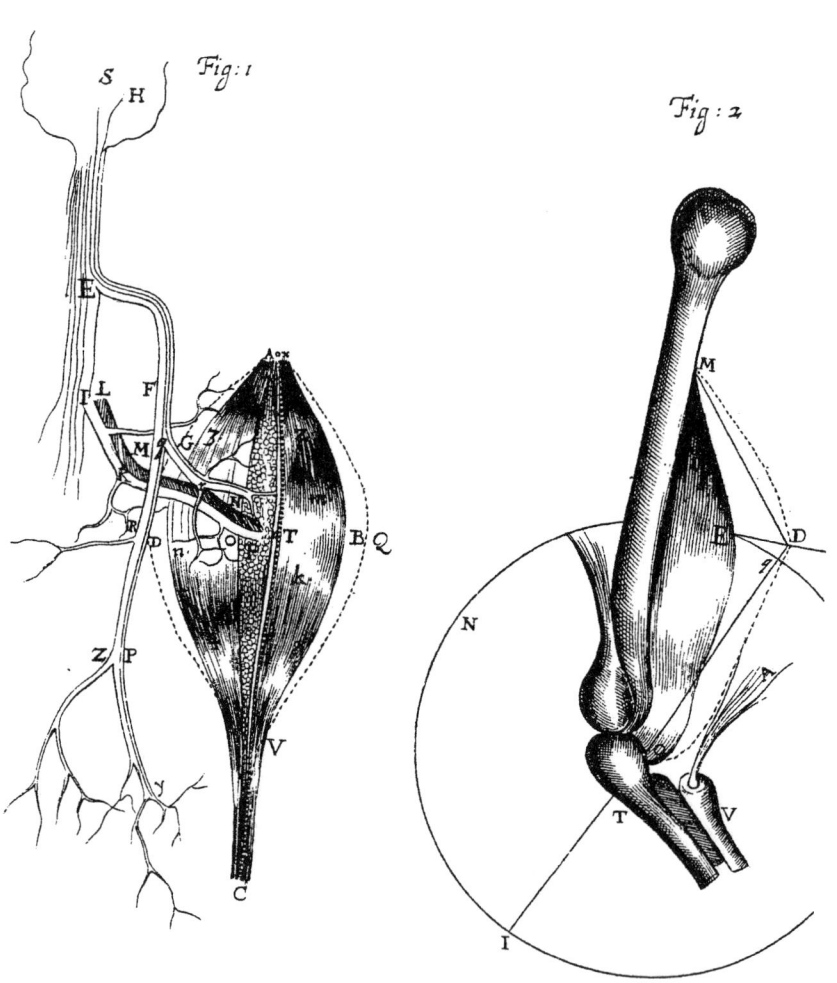

1

D E

Ratione Motus Musculorum.

1. DE motu Musculorum acturus non necessum esse arbitror, ut ipsorum fabricam ac Constructionem hoc in loco fusiùs enarrem: Quippe quæ jamdudum à pluribus Anatomicis, ac præcipuè à Divino illo Vesalio, ac Hieron. Fabr. ab Aquapendente, clarè copioséq; tradita sit. Quantum verò ad institutum meum opus fuerit, suo ubiq; loco interseram; Nihil tamen adducturus sum nisi quod optimorum Anatomicorum fide, aut saltem nostrâ qualicunq; observatione, exploratum perspectumq; sit.

2. Duæ ergò duntaxat, quod sciam, hactenus obtinuerunt apud Medicos ac Philosophos, de motu Musculorum, sententiæ; Altera, Ipsos moveri à spiritibus (quos propterea vocant Animales) in cerebro quidem elaboratis, deinde ad Imperium animæ seu voluntatis, per Nervorum canales in singulos Musculos traductis, decernit: Sed alii hanc rem aliter explicant. Vett: enim cum Galeno, Motricem nescio-quam facultatem cùm spiritibus hisce vectoribus per Nervos in Musculum deferri volunt; Musculum autem continuò spiritibus his Illustratum (ut verbis utar Bauhini) obtemperare, partemq; pro Imperio voluntatis movere: de quâ hoc tantum dixerim, Omnino me ipsam non intelligere.

3. Recentiores autem, uti vir ille Incomparabilis Renatus Cartesius, & ex ipso Cl. V. Henricus Regius, &

B   post

On the reason
of the movement of the muscles

1. I think that in dealing with the movement of the muscles it is not necessary that I describe here at length their anatomy and their structure. These things, of course, have long ago been made abundantly clear by many anatomists, and especially by the excellent Vesalius as well as by Hieronymus Fabricius of Aquapendente. In the appropriate places I shall mention what will be needed for my purpose. I shall, however, bring forward only that which has been investigated and is well known according to the best anatomists or at least from our observation of whatever kind.

2. As far as I know, two opinions have prevailed hitherto among physicians and philosophers concerning the movement of muscles. One declares that the muscles are moved by spirits elaborated in the brain (and called for that reason animal spirits), which are then carried to the different muscles through the canals of the nerves, following an order of the mind or of the will. But others explain this matter in another way. The ancients indeed, with Galen, claim that some motor faculty or other is carried through the nerves into the muscle with these spirits as carriers; and the muscle, continuously irrigated by these spirits, complies (to use the words of Bauhin) and moves the part at the order of the will. I shall only say that I do not understand at all this point of view.

3. More recent authorities, however, such as the incomparable René Descartes, and after him the very famous Henry Regius, and

post hos Alii, hanc fpirituum per Nervos tranfmiffionem
longe alia ratione exponunt: quorum quidem opinionem
per partes impræfens non deducam, quoniam jam omnibus fatis innotuit.

4. Altera verò fententia illorum eft, qui fic omnes
Mufculos partibus movendis Attenfos effe dicunt, ut eorum Fibræ in perpetua quadam, ac forti Tenfione confti-
tuantur; ac proinde quidem femper quilibet Mufculus
fe naturali fponte contrahat, quotiefcunq; per virtutem,
Nervi interventu, è cerebro delapfam eum vigorem nactus
fit, ut vim fui Antagoniftæ fuperare poffit. Et in his quidem eft D. D. Scarborough, fummi Ingenii ac Eruditionis
Medicus.

5. Quid in utraq; fententiâ defiderari poffe videatur,
priùs veniam præfatus, breviter edifferam: Ac ifta profectò Cartefii, ut ab ipfo in Pofthumo illo de Homine
tractatu explicatur, multa in fe præclara habet, & omnis
veræ Philofophiæ Principis Ingenio dignifima: quia tamen haud pauca affumit, quæ ipfâ rerum fide minùs conftare exiftimo, (uti illud, in Nervis Valvulas quafdam ineffe, deinde & fpiritus animales Venti inftar habere, ac ab
ipfis violente irruentibus Mufculos, veli ad modum, inftari
ac diftendi,) igitur huic affentiri nequeo: Præterquam
enim, quòd iftiufmodi valvulas nulla hactenus Profectorum
fubtilitas affequi potuit, Quis, obfecro, eas in nervo
intelligat, qui, uti fentit Cartefius, ipfaq; Anatomia te-
ftatur, ex infinitis quafi funiculis filamentifq; communi
membranâ inclufis, contexitur? Quòd fi porrò Mufculi
fabricam fpectemus, Nerviq; in ipfum inferti exilitatem,
cùm ejufdem mole, carneâ, nullifq; cavitatibus donatâ,
conferamus, neq; ipfum fpiritibus è nervo exiguo quafi
flatu quodam itâ diftendi poffe, neq; adeò quicquam cùm

velo

others after them explain this transmission of spirits through the nerves by a far different reasoning: I will not bring out their opinion in detail now since it is sufficiently known by everybody.

4. The second opinion is that of those who say that all muscles are attached to parts to be moved so that their fibers are set in some permanent and strong tension. Consequently, any muscle always contracts spontaneously whenever, by a strength which comes down from the brain by way of a nerve, it receives so much vigour that it can overcome the force of its antagonist. The very learned Dr Scarborough,[1] a physician of the utmost talents and erudition, is among those who hold this opinion.

5. Let me first be permitted to set forth briefly what seems to be lacking in each of these two opinions. Indeed that of Descartes, as presented by himself in his posthumous *Treatise on Man,* contains much that is remarkable and worthy of the genius of a prince of true philosophy. However, it assumes many points which I think do not conform with reality (for example that there are some valvules in the nerves, and then that animal spirits behave like wind and that the muscles are inflated and distended like a sail by their violent irruption). I cannot therefore agree with this. Besides, indeed, the fact that no skill on the part of dissectors has been able to find valvules of this kind, who, I ask, will understand them in a nerve which is made up of an infinite number of some sort of threads enclosed in a common membrane, as perceived by Descartes and shown by anatomy itself? Moreover, if we consider the structure of a muscle and if we compare the thinness of the nerve inserted in it with its mass of flesh devoid of any cavities, we are led to think that a muscle cannot be thus distended by spirits emerging as it were in a sort of puff from a slender nerve, nor does it have anything

[1]This is probably Sir Charles Scarburgh., according to Nayler and also Wilson.

velo commune, habere advertemus.

6. Quod ad alteram nunc Sententiam attinet, puto equidem eam partim esse veram, multisq; experimentis confirmari: Enimverò, si principium Musculi (ut à Galeno olim,& recèns à Clariss. illo Anatom. Aquapendente, observatum est) supernum abscideris, totus ad caudam fertur; si mediam, versus utramq; extremitatem conglobitur; si finem Inferiorem, ad caput retrahitur: Adhæc, Musculus excarnis (ut idem ille annotavit) Naturalem quandam contractionem perpetuò affectat. Deniq; & illud accedit, quòd si cui, vel invito maximè ac renitenti, brachium flectas, Musculi superiores, ceu cùm illud propriâ sponte incurvamus, tumescunt: hoc autem spiritibus animalibus, qui tantùm ex mandato Animæ è Cerebro exire creduntur, minimè attribuas.

7. Cùm verò ipsi hujus opinionis autores, Musculorum in corpore partem longè maximam, suos habere Antagonistas, viribus sibi æquipollentes, fateantur; certè, vim illam Insignem esse oportet, quæ æquale in Antagonistâ Musculo robur pervincat; atqui, ea nondum ab ipsis explicata est: Veruntamen sic illam explicari posse spero, & ad leges etiam Mechanicas exigi, ut nobis aliam plane contrahendi Musculi causam, hac ipsâ quidem si non potiorem, saltem in agendo priorem, suppeditet: Interim autem cùm Galeno, qui virtutem illam in Musculis sponte se contrahendi primus annotavit, confiteor, Hanc quoq; unà cum istis, ad motum Musculi perficiendum, operam non levem adferre: nam per se non sufficere, ipse Galenus, eodem hoc argumento usus, contendit.

8. Principiò autem illud ostendam, Omnem Musculorum motum à spirituoso quodam liquore è nervis exstillante administrari; eâ tamen ratione, ut motus iste hinc

in common with a sail.

6. As far as the other opinion is concerned, I think it is partly true and it is confirmed by many experiments. If indeed (as was observed formerly by Galen and recently by the famous anatomist Aquapendente) you divide the upper end of a muscle, all the muscle moves towards its lower end. If you divide the muscle through its middle, it retracts towards both ends. If you divide its lower end, it retracts towards its upper origin. Besides, a resected muscle (as Galen himself noticed) permanently presents with some natural contraction. Finally it also happens that, if you flex the arm of somebody entirely against his will and against his resistance, the upper muscles swell just as when we flex the arm willingly.[2] You cannot assign this to animal spirits which are believed to go out from the brain only at the command of the mind.

7. The authors who hold this opinion claim that by far the greatest part of the muscles in the body have their antagonists which are equal in force to them. Certainly, this force must be considerable which overcomes an equal force in an antagonistic muscle. But yet this force has not been explained by these authors. Nevertheless I hope to be able to explain this force and to show that it complies with the laws of mechanics, so that it provides us with a completely different cause of muscle contraction, that cause being, if not the more important one, at least the one that is earlier to act. But meanwhile, with Galen who was the first to notice this power of the muscles to contract spontaneously, I recognize that this power together with those other powers contributes greatly to the movement of the muscle. Indeed Galen himself, using the same reasoning, asserted that it was not sufficient by itself.

8. But, to begin with, let me show that every movement of muscles is directed by some spirituous liquor trickling from the nerves, in such a way, however, that the movement is only

---

[2]See Introduction, p. 19 and Fig. 3, p. 20 .

4

folummodo primùm inchoetur; deinceps verò, ab aliis
duabus caufis admodùm valentibus poftea abfolvatur.
Quod dum facio, mihi hic nonnulla paulò altiùs repeten-
da funt.

Nolo autem me fruftrà torquere, in modo illo quo Anima agit in Corpus inveftigando: quem nemo forté mor-
talium intelligendo unquam confequi valebit: Profectò
ità fe rem habere, vix credo quifquam eft qui dubitet:
Attamen non ab ipfa immediate, verùm intercedente aliquo inftrumento materiali, grandes illas Mufculorum Ma-
chinas tractari; id quidem ex eo manifeftum eft, Quòd in
Convulfionibus omnes illorum motus, abfq; animæ ope,
præftari videamus: nifi etiam, Animam ipfam à Morbo
affici, ac in rabiem quandam agi putes, unde tam incle-
menter corpus fuum diftrahat, ac furialibus modis con-
vellat.

9. Quinetiam illud quicquid fit à quo omnis Mufculus
fuum motum orditur, per Nervorum certè canales traduci
necellum eft: Námq; iis incifis vulneratifq;, aut demùm
validè ligatis, fenfus, motúfve, aut quidem uterq;, conti-
nenter Mufculo perit: Præterea, fi nervum fecueris, pars
Mufculi fuperior verfùs cerebrum, fenfum motúmq; reti-
net; & vice verfa: fin verò medulla Spinalis, unde nervi
quamplurimi in varias corporis partes educuntur, fecta fit;
nervi infrà fectionem orti fenfum ac motum amittunt, alii
non item. Ex quibus utiq; neceffariò confequitur, Spon-
taneam illam fibrarum Mufculi contractionem nihil effi-
cere, nifi porrò vis illa, quæ tandem cumq; fit, è cerebro
per Nervos accefferit. Unde merito cùm Galeno admi-
rari licet, Quemadmodum nervus, mole quidem exiguus,
tam immenfà virtute præditus fit.

10. His infuper . a quoq; adjcienda funt; Quòd enim
                                                         q;ilfq;

initiated by this; it is then completed afterwards by two other very strong causes. In so doing, I have here to return to some things I mentioned earlier.

But I do not wish to torment myself in vain about how the mind acts on the body. Nobody perhaps amongst the mortals will ever succeed in understanding this. Assuredly I believe that the truth of the matter is like this: these big machines of the muscles are not activated immediately by the mind but through the intervention of some material instrument. This is obvious from the fact that in convulsions all the movements of the muscles seem to be carried out without need of the mind, unless you think that the mind itself is affected by the disease and is driven into a kind of frenzy whereby it distracts its body so harshly and contorts it furiously.

9. But indeed that, whatever it is, by which every muscle starts moving, must certainly be transmitted through the canals of the nerves. Indeed if these are cut into and injured or tightly ligated, sensation or movement or sometimes both are lost as far as the muscle is concerned. Moreover if you have cut through a nerve, the proximal part of the muscle, towards the brain, retains sensation and movement and vice-versa[3]: but if, indeed, the spinal cord from which most nerves are led forth to the different parts of the body, is cut through, the nerves originating below the section are deprived of sensation and movement, not the others. From these facts assuredly it results necessarily that the spontaneous contraction of the muscle fibers accomplishes nothing unless aferwards that force, whatever it may turn out to be, arrives from the brain through the nerves. Therefore, one can deservedly wonder with Galen how a nerve, small as it is in volume, is endowed with such immense power.

10. Beside these facts, there are also others to be considered.

---

[3] It is the muscles above the site of the lesion which retain sensation and movement, not the upper part of the muscle (note of the editor).

quifq; Musculus est mole amplior, usúq; frequentior, fortiórq;, eò majorem Nervum sortitus est; & è contrá. Deinde, illud maximè annotare convenit, Nervos ut plurimùm in Musculi capita ac principia Implantari, deorsúmq; spargi: nonnulli quidem in mediam eorum partem se insinuant, semper autem ad interiora feruntur: Nullus unquam, Galeno dicente, in finem inseritur: Quippe Musculi omnes sese versùs caput solummodò contrahunt. Fateor equidem, in longissimis quibusdam, ubi Nervus Principio immissus per totam ejus longitudinem commodè duci nequibat, alium interdum qui Cerebro aut Spinali medullâ propinquior sit, ex partibus vicinis quasi in subsidium venire, Musculúmq; versùs finem perforare; ità tamen, ut nullæ ejus divaricationes superiora petant. Ex quâ profectò Nervorum in Musculum Immersione, apertè colligitur, Ipsos iis movendis inservire. Quod idcircò monere placuit, quoniam eandem planè Arteriarum Venarúmq; rationem esse, paulò infrà videbimus.

11. Cùm ergò tam luculenter constiterit, vim quandam per nervos in Musculum deferri, absq; quâ, ipse quidem ad motum inhabilis ac ineptus sit; restat, ut cujusmodi jam illa sit, breviter inquiramus: Si nervum itaque attentiùs contemplemur, omnem ipsius fabricam, è substantia quadam medullari, succo pertusâ; membrana duplici, quæ medullam istam involvunt; ac infinitis demùm funiculis intrà membranas hasce medullámq; à principio ejusdem usq; ad extrema capillamenta extensis, constari animadvertemus: præter hæc autem, nihil omninò, quod sciam, in Nervis conspicitur: Deinde etiam paulispe. cogitemus, Quemadmodum hi nervi intrà Musculos de? seminentur; etenim ceu Caudex herbæ in minores ser. ac minores ramulos diffisi, tandem in Flexus qui

For, indeed, the bulkier, the more frequently used and the stronger any muscle is, the bigger the nerve it is assigned, and conversely. Furthermore, it is appropriate to notice this especially: nerves are implanted for the most part into the heads and beginnings of muscles and spread out downwards. Some nerves are inserted into the middle part of the muscle but they are always carried to the lower parts. None, Galen says, is ever inserted into the end. Of course all the muscles contract only towards their head. I admit that, in some very long muscles, where the nerve inserted at the beginning could not be led appropriately throughout its whole length, another nerve which is closer to the brain or the spinal cord, occasionally comes from the neighboring parts as if it were auxiliary, and perforates the muscle towards its extremity, but so that none of its branches reach into the upper parts. From this insertion of the nerves into the muscle, it can be obviously deduced that the nerves serve to move the muscles. This I decided to point out since we shall see a little further on that the system of the arteries and veins is the same.

11. As it has been so clearly established that some force is carried through the nerves into the muscle, without which the muscle is incapable and unsuited to movement, we have still to examine briefly the nature of this force. If we observe a nerve more carefully, we will notice that its structure is all made of a medullar substance perfused by a juice, two membranes which envelop that medulla itself, and finally an infinity of small threads extending inside these membranes and the medulla from the beginning of the nerve to the extremity of the smallest ramification. Besides these elements, however, absolutely nothing else is observed in nerves, as far as I know. Then let us think a little and consider how these nerves are disseminated inside the muscles. Like the stem of a plant, they divide into smaller and smaller branches and finally end in some

6

Membranofos abeunt, in quibus omnino difpenduntur &
evanefcunt.

12. Quo autem fequentibus lucem majorem afferamus,
pauca hic de naturâ fpirituum (fic enim, vocabulo à Chymicis accepto, Particulas fubtiliores & actuofas iftorum li-
quorum quibus corpus irrigatur, rectiffimè appellant) di-
cenda funt. Notum eft itaq;, Succum alimentarem particulis quibufdam fubtilioribus refertum effe, quæ per fre-
quentem fanguinis, cui admifcetur, circuitum, paulatim è
craffis ac terreftribus quibus implicitæ funt, evolvuntur;
hoc eft, ut appellationibus Chymicis etiamnum utar, è
ft itu Fixationis ad ftatum Volatilitatis perveniunt.

Hujufmodi Particulæ in fanguine arteriofo admodùm
abundant, licet cùm maximâ aliarum copiâ permiftæ:
dum verò is per arterias Cerebri fertur, liquorem ex fe
Mercurialem, h. e. fale ac fulphure volatili exquifitè im-
prægnatum, lentâ quadam deftillatione in cerebri medullam deponit, qui exinde in omnes totius corporis Nervos
depluit: per hos autem pedetentim incedens quaquaversùm difpenfatur, ac tandem tardiori circulatione in venas
effunditur, denuóq; ad cor redit. Spirituofis hujufce generis
liquoribus, fingulæ in Animantis corpore partes vehementer turgent; etfi pro variâ partis temperie, ac fertmeni
in ipfâ peculiaris ratione, varii fint. Omnes autem ubicunq;
funt, caloris nativi ac circulationis beneficio, in conftanti
motu ac agitatione exiftunt: & hæc ipfa Agitatio, id ipfum eft quod Vitam appellamus.

13. Qui materiam hanc Spiritûs nomine indigitârunt,
eam fub formâ halitûs cujufdam aut venti exiftere voluére;
adeóq; Mufculorum Intumefcentiam ac Contractionem,
à velociffimâ ejus inflatione proficifci: Atqui id certè nal-
lo prorsùs idoneo argumento innititur. Ubinam enim, obsecro,

membranous folds in which they are entirely lost and disappear.

12. In order to shed more light on subsequent points, however, there are a few things which should be said here about the nature of the spirits (by which term, taken from the chemists, people rightly name the very subtle and lively particles of those liquors by which the body is irrigated). Thus it is well known that the alimentary juice is crammed with some subtle particles which, through the frequent circuit of the blood with which this juice is mixed, are progressively extricated from the gross and earthy particles with which they are entangled. This means, to use once more the terminology of the chemists, that they pass from a state of fixation to a state of volatility.

Particles of this kind abound in the arterial blood, although they are mixed with a huge quantity of others. While being carried through the arteries of the brain, the blood, by some sort of slow distillation, secretes a mercurial liquor, i.e. impregnated finely by salt and volatile sulphur, into the medulla of the brain. From there this liquor then drips down into all the nerves of the whole body through which, proceeding little by little, it is dispersed everywhere and finally, circulating more slowly, it is poured into the veins and returns to the heart. The separate parts in the body of an animal swell impetuously by the effect of spirituous liquors of this kind, although they vary according to the different character of the part and the proportion of a peculiar ferment in this part. But all of them, wherever they are, thanks to their innate heat and to the circulation, are in constant movement and agitation. This agitation itself is exactly what we call life.

13. Those who have designated this matter with the name of spirit claimed that it exists in the form of some exhalation or wind and that, therefore, the swelling and contraction of the muscles proceed from its being blown in very swiftly. But this certainly does not by any means rest on any satisfactory line of reasoning. Indeed where on earth,

7

fiero, ad hafitus hofce encipiendos fpecus excavantur, aut,
Unde tanta eorum copia, ut Mufculos, nulla cavitate præ-
citos, tam fubitò inflent? Quod fibi faciendum putavit
Magnus nofter Harvæus de fpiritibus vitalibus agens, idem
quoq; ipfe, confifus oculorum teftimonio, fecerim ; Nega-
vit quippe, Tales extrà fanguinem, dum, quòd eos in diffe-
ctionibus nufpiam repererit. Quoties igitur Spiritus Animales dico, liquorem iftum Nervorum fubtiliffimum, activ-
offimum, fumméq; volatilem, intelligo ; planè ac cùm
fpiritum vini, aut falis, aut alios hujus generis, appellamus.
Quamobrem fpiritus in Nervis Animales vocatos, nihil
aliud effe præter reci ficatum hujufmodi ac prædivitem
fuccum, planè perfuafum habeo: cùm ipfa doctina con-
teftetur, Talem omninò effe nervum qualem fuprà defcri-
pfimus ; non autem Tubulorum ritu excavatum, ut alte-
rius fententiæ autores affirmare videntur: Nimius fim, fi
ea hîc recenferem, quibus Vir Cl. ac Eruditiff. Gliffonius,
Liquorem hunc quem loquor fpirituofiffimum continenti
circuitu per nervos deferri adftruxit. Illud faltem attige-
rim, In omni Mufculo fpiritus non unius generis ineffe:
alius enim eft in Tendine ejúfq; fibris, alius in carne Mufcu-
lofâ, alius demùm qui per nervos affluit.

14. Si ea quæ jam dicta funt, cùm iis quæ modò diffe-
rebamus conferantur, omninò arbitror, Animam fpirituofos
hofce liquores, ad omnes tam fenfionis quàm motus actio-
nes præftandas adhibere. Quod inde etiam ampliùs confirmatur, quòd à vini aut alicujus alius liquoris fpirituofi lar-
giori hauftu, omnes corporis motus multò agiliùs obiri videamus: contrà autem laffitudo five à morbo five ab ex-
ercitio vehementi, non aliunde quàm à defectu fpirituum
profeifcitur. Reliquos autem ufus, quibus prætereà fervi-
unt, impræfens miffos facio.

15. At

I ask, are there cavities hollowed out to receive these exhalations and from where do they come in such great quantities that they inflate so suddenly the muscles which are not provided with any cavity? Relying on the testimony of my eyes, I will myself do what our great Harvey held that he must do when dealing with vital spirits. He denied that there are such things outside the blood since he found them nowhere in his dissections. Thus, whenever I speak of animal spirits, I mean this very subtle liquor of the nerves, very lively and extremely volatile, exactly as we speak of spirit of wine or of salt or of others of the same kind. Therefore, I am quite convinced that what is called animal spirits in the nerves is nothing else than a very rich and rectified liquor of this kind. As autopsy bears witness, a nerve is absolutely such as described above, not hollow as tubules are, contrary to what authors of the other opinion seem to assert. It would be more than is needed if I listed here the reasons on which the famous and very erudite Glisson based [the concept] that this liquor which I call very spirituous is carried through the nerves in a continuous circuit. Let me at least mention this: in the whole muscle there is no spirit of one kind alone. The spirit in a tendon and its fibers, that in the flesh of the muscles, and finally that which flows in through the nerves are all different.

14. If what has been said so far is brought together with what we just discussed, I definitely think that the mind employs these spirituous liquors to carry out all functions of sensation as well as of movement. This is even more strongly established by the observation that all movements of the body are carried out much more quickly after drinking too much wine or some other spirituous liquor. On the contrary, tiredness either from disease or from strenuous activity results only from a lack of spirits. But for the moment I pass over the other functions which they serve.

B

15. At quoniam Sensionis explicatio non parum ad ea
quæ dicturus sum illustranda conducit, obiter indicabo in
quo ea mihi consistere videtur : Nè itaq; longum faciam,
puto Liquorem hunc rectificatissimum, summéq; volatilem,
prout in membranis corporis hospitatur , una cum istis
membranis, sensús esse Instrumentum ; quatenùs autem
in Cerebri ac Nervorum medullá est, motui producendo
inservire: Utriúsq; rationem non aliunde quàm ab experimentis ducendam censeo: Sive ergò receptissimam Anatomicorum sententiam , sive quod potius est, omnem
Medicorum ac Chirurgorum praxin excutiamus ; certè ex
omnibus corporis partibus , unicam esse Membranam ,
quod immediatum Tactus organum sit , comperiemus.
Etenim Medullam cerebri Sensús instrumentum non esse,
ex eo liquet , quòd frustula ejus insigni magnitudine ,
absq; ullo dolore, in magnis Capitis vulneribus exempta
fuerint: in aliquibus etiam, quibus sensus superstes erat,
aut in lapillos per partes indurata, aut tota in lapidem versa est, ut bovi isti Suecico, de quo Cl. Bartholin, in Histor.
Anatom. Cent.6. contigit ; Jure profectò mirabatur Vir
Cl. unde ipsi motus remanserit, & quod ipse optimè sus-
picatus est, sententiam nostram præclarè confirmat ; erant
enim in lapideo isto cerebro foraminula hinc & inde, in
quæ paleæ immitti possent, exculpta, per quæ spiritus Animalis, ex arteriis in Nervorum medullam liberè comme-
aret. Quibus accedit, in Nervorum vulneribus, ipsorum
pariter medullam absq; ullo omninò cruciatu innoxiè con-
trectari ; in membranis autem, sensus exquisitissimus ha-
betur.

16. Existimo autem istos Anatomicos rectissimè sentire,
qui omnes in corpore Membranas sensiles, à cerebri Tunicis oriri volunt ; quod ab ipsá, quantum videre possum,
autopsiá

15. And because an explanation of sensation helps considerably to throw light on what I am about to say, I shall indicate in passing what sensation appears to me to consist of. To be brief, I think that this very rectified and extremely volatile liquor, in so far as it is lodged in the membranes of the body, is, together with these membranes, the instrument of sensation. But in so far as it is present in the medulla of the brain and of the nerves, it serves to produce movement. I think that the understanding of both functions must be deduced from nothing else than experience. Whether therefore we examine the most accepted opinion of the anatomists or rather that which is the whole practice of physicians and surgeons, we will certainly find out that, among all the parts of the body, there is only one membrane which is the immediate organ of [the sense of] touch. It is clear indeed that the medulla of the brain is not the organ of sensation since, in large injuries of the head, considerable pieces of it have been removed without causing any pain. In some [animals] even, where sensation survived, the medulla was either partially hardened into little stones or was completely changed into stone, as in that ox of Succico of which the very famous Bartholin speaks in *Histor. Anat. Cent.*6 where the famous man rightly wondered why movement remained in the ox, and what he had very well surmised clearly confirms our opinion. In this brain turned into stone, there were indeed small holes here and there into which straws could be inserted and through which animal spirit would flow freely from the arteries into the medulla of the nerves. Moreover, in injuries to nerves, their medulla is similarly touched without any pain; but in the membranes there is the most exquisite sensation.

16. I thus think that those anatomists who claim that all sensitive membranes in the body originate from the tunics of the brain, are absolutely right. As far as I can see, this is confirmed by

autopfia comprobatur : unde patet, quomodo iftius membranæ beneficio qui omnibus totius corporis Mufculis ob-
tenfa eft, ill úfq quæ fingulos privatim Mufculos inveftit,
ac demùm nervorum qui per omnes fparguntur, univerfis
ferè partibus fenfus attribuatur. Præterea, illud valdè an-
notandum eft, Omnes iftas corporis membranas, quod
fuprà de Mufculis dicebatur, in fummâ quoq. Tenfione
exiftere: ac etiam liquore quodam proprio admodum
fubtili, five fpiritu vivifico, impleri, qui per omnes earum
fibras perpetuò tranfit, ipfafq; in debito Tono confervat.
Nec quifquam hoc nimium effe ad concedendum putabit,
qui minutula illa animalcula omni fuo apparatu inftructa
cogitat, vel exiliffimis termè harum fibrillis minora.

17. Intelligendum verò eft ; Membranas hafce, dum
Tenfionem fuam ac Tonum debitum obtinent, videri
poffe fe ad modum Campanæ aut Vitri puriffimi habere,
quorum ea eft indoles, ut, fi partem aliquam ferias, reliquæ
univerfæ, tremulâ quadam vibratione concutiantur ; &
hoc pacto, omnia fenfuum objecta, vel interceffu membranæ iftius Nervi qui ad certum alicujus fensûs organum
fpectat, vel communis illius qui totum corpus involvit,
per lineas, quantum fieri poteft, rectas, ad cerebrum deferri autumo: in quo varii diftinctiq; objectorum motus
ab animâ percipiuntur, quod quomodo fiat infrà Sect.16.
clariùs dicam. Hinc etiam mirus ille nervofarum partium
confenfus oritur: ac fortaffè ex his lucem quoq; aliquam
perdifficili ifti quæftioni foenerabimur, Cur fenfus pere-
unte motu, in Paralyfi fuperfit, & contra. Namq; fi ifte
Membranarum Tonus, vel univerfim, vel ex parte cor-
rumpatur, ac naturalis, qui fuit, particularum fitus ac or-
do turbetur, vel accedente nimia humiditate, vel ex vulnere, aut alio quovis modo ; tum quidem vibratio illa ac
                    (unculae)

autopsy. This shows how sensation is attributed to almost all the parts of the body thanks to this membrane which extends to all the muscles of the whole body, and to that which envelops the different muscles separately as well as to the nerves which are dispersed through all the [muscles]. Moreover, it is very noteworthy that these membranes of the body are also under the greatest tension, as was said above about the muscles. They are also full of some characteristic and very subtle liquor or lively spirit, which continuously passes through all their fibers and maintains them in due tone. Nobody will think this is too much to admit, who considers all those very small animalcules provided with all their equipment, even smaller than the very smallest fibrils of these [membranes].

17. It must be understood that these membranes while they maintain their tension and due tone can be considered as behaving like a bell or as the purest glass. These have the property that, if you strike any part, all the remaining [parts] are shaken by some sort of quivering vibration. In the same way, I say that all objects of the senses are carried to the brain by the intervention of either the membrane of the nerve which pertains to the true organ of that sense, or of the common membrane which involves the whole body, carried in lines as straight as possible. In the brain the different and distinct movements of the objects are perceived by the mind. I will explain more clearly in section 26 below how this occurs. This is the origin of the marvellous harmony of the nervous parts. Perhaps from these things we shall also shed some light on this very difficult question: why does sensation persist in paralysis after the movement is lost and conversely? If indeed that tone of the membranes is totally or partially destroyed, and if the normal situation and order of the particles, which existed, is disturbed, either by the accession of too much moisture, or by an injury, or in any other way, then that vibration and

10

undulatio particularum, ex quâ fenfus fit, intermumpitur : ut in Campanâ aut Vitro, cui fiffura accidit, fonum illum Tinnulum ac acutum, in raucum quendam & ingratum ftrepitum permutari notamus. Contrà, cùm al'æfis membranis, extraneo quodam fucco medulla dæffait ac vitiatur, illicò motus perit, quem à generofo illo liquore in nerv.s, quo medulla ipfarum continuò irrigatur , produci jam proximè demonftrare conabimur.   Quòd autem Tonus quidam ac vigor tum fubftantiæ medullaris Nervorum, tum etiam in membranis ad fenfum ac motum requiratur, ex Somno patere videtur : in quo latex hic fpirituofus per nervos membranafq; interiores ac majores tantummodò (ut ipfe fanguis per vafa majora, unde dormientes pallent) fuum circuitum perficere exiftimandus eft.   Hinc Tonum debitum minoribus ac exterioribus ad fenfum motumq; efficiendum præftare nequit ; ac adeò fi oculos dormienti aperias non videt, ob nimium Membranarum laxitatem ac flaccefcentiam.

18. Ex iis quæ fuperiùs dicta funt, abundè conftat, Aliquid in Mufculum deduci, abfq; quo nullus ipfi motus conficitur: effe verò illud materiam quandam fummè mobilem tum per ipfos nervos, tum etiam per Mufculorum fibras decurrentem, præter ea quæ jam ad hoc probandum allata funt, etiam ad oculum oftendi poteft.   In mactatis enim recèns Animalibus, cernes, detracto corio, materiam hanc, five Spiritus, quos vocant, tumultuariò curfantes, exteriorem Mufculofæ carnis fuperficiem variis modis fecundùm omnes fibrarum ductus pertrahere: Effe verò illius motum celerrimum tum ex hoc quod jam commemini, tum alio quoq; experimento, manifeftum fit ; quoties enim Cubitum fortè ad duram aliquam materiam ex improvifo allidimus, Tremor quidam feu vibratio dolorifica, ab ipfo
cubito

oscillation of the particles which make the sensation is interrupted. In a bell or a glass in which a crack occurs, we notice that the tinkling and high-pitched sound is altered into a harsh and unpleasant clanking. On the other hand, when the membranes are intact, if the medulla is dissolved in and vitiated by some extraneous juice, movement immediately disappears. We shall presently attempt to show that this movement is produced by that excellent liquor in the nerves, by which their medulla is continuously irrigated. Sleep seems to make clear that some tone and vigor is required both of the medullary substance of the nerves and in the membranes for sensation and movement. In sleep it must be reckoned that this spirituous fluid carries out its circulation through only the inner and larger nerves and membranes (as blood itself flows only through the larger vessels for which reason those who are asleep turn pale). Hence, the fluid is not able to produce in the smaller and outer [nerves] the tone necessary for achieving sensation and movement. Therefore, if you open the eyes of somebody asleep, he does not see, as a result of too much laxity and flaccidity of the membranes.

18. From what was said above it appears clearly that something is brought into the muscle, without which no movement of the muscle is accomplished. Besides what has been presented already to commend this, visible proof can be given that this is some extremely mobile substance coursing down through the nerves and also through the fibers of the muscles. Indeed in animals recently slaughtered, after the skin has been removed, you discern this substance, or spirits as people call it, running disorderly [and] deforming the outer surface of the muscular flesh in various ways, following the lines of the fibers. The movement of this substance is very swift. This is obvious as well from what I have just mentioned as also from another experiment. Every time indeed that we unexpectedly chance to knock an elbow against some hard material, a tremor or painful vibration flies from the elbow down

35

cubito ad extremos ufq; digitos quafi ictu oculi pervolat:
Confimiliter, ac cùm bacillus è ferro longior, in alterâ
extremitate percuffus, tremulam omnium particularum
vibrationem edit, quam, manu alteri admotâ, facilè
fentias.

19. Quò hæc autem rectiùs intelligantur. pauca de
Mufculi conftructione prælibanda funt.   Is ergò ex infinitis fibris Tendinofis quafi chordis conflatur: quæ in utroq;
ejus extremo coeuntes irà intorquentur & coalefcunt invicem, ut funem aliquem grandiorem, ex innumeris aliis
minoribus textum referant: intra verò Mufculi corpus,
ceu ftamen in fila difcerptum à fe invicem diftant; Omnia
autem inter ipfas fpatia ac intervalla carne ubiq; inferci-
untur.   Quibus fi Membranas duas, Mufculum Tegentes,
ac ómnes fibrillis ex fe emiffis fibi invicem colligantes,
fingulofq; ab aliis difparantes, ac etiam exquifitâ fuâ Ir-
vitate fummam movendi facilitatem ipfis tribuentes addi-
deris, ac deniq; vafa in ipfum derivata, Nervos, Arterias,
Venas, Lymphæductus annumeres, Integrum Mufculum
cernis.

20. De Fibris autem Tendinofis tria fummoperè no-
tanda funt: Primò, ex iis potiffimum Mufculos conftare:
quòd ex eo liquet, quòd Octaginta librarum Pondo alli-
gatum iftius Mufculi Tendini, quem Gracilem Internum
in homine vocant, ab humo fublatum facilè fuftinuerim,
alterâ Mufculi extremitate manu apprehensâ. Nec dubito
quin multò majus Pondus, fi ad manum fuiffet, fuftulif-
fem; nifi verò totus ferè ex firmiffimis hifce fibris confti-
tiffet, tantum ponderis nunquam extollere potuiffet: Taceo hoc idem quoq; in Mufculo excarnato clariffimè cerni;
adeo ut nullos Mufculos, quicquid dicant aliqui, tendine
fuo, divifo faltem ac in fibras foluto careat.

to the tips of the fingers as it were in the twinkling of an eye. Similarly, when a long rod of iron is struck at one end, it produces a quivering vibration of all the particles which you feel with your hand at the other end.

19. Better to understand these matters, a little must be said about the structure of a muscle. This is made up of an infinity of tendinous fibers almost like strings. Converging at both ends of the muscle they are entwined and coalesce so that they form a larger rope made of countless other smaller ones. Inside the body of the muscle, like a string unraveled into threads, they are separate from each other. But all the spaces and intervals between them are filled everywhere with flesh. If you add to these the two membranes coating the muscle and connected to each other by all the fibrils originating from them, and the individual fibrils separate from the others, and if you impart to them an extreme facility to move thanks to their exquisite lightness, and finally if you take into account the vessels which penetrate them, nerves, arteries, veins [and] lymph ducts, you have an idea of a whole muscle.

20. But as far as the tendinous fibers are concerned, three points must be noticed above all. Firstly, the muscles are made principally of them. This appears from the fact that I have easily maintained, lifted up from the ground, a weight of eighty pounds attached to the tendon of the muscle which in man is called gracilis internus, the other extremity of the muscle being held in my hand. I have no doubt that I should have supported a much heavier weight, if one had been at hand. Truly, it could never have raised such a weight if almost all the muscle had not consisted of these very strong fibers. I say nothing about the fact that this is also very clearly observed in a muscle removed from the body. This is such that no muscle, whatever some people say, lacks its own tendon, divided at the least and separated into fibers.

83

21. Secundò, Hæ de quibus agimus Fibræ, non modò pro ratione fitûs partium moventarum varios ductus obtinent, veruntamen etiam, quod à nemine hactenus satis observatum est, Fibræ ejusdem Musculi, iisdemq; motibus destinatæ, variis omninò implicationibus contorquentur, ac sibi invicem interseruntur: ut in semi-nervoso corporis humani videre est, quem veluti ex aliis intrà se inclusis Musculis constari putes. In aliquibus, fibræ rectâ à capite ad caudam incedunt; in aliis, plus minúsve, per lineas quidem spirales aut wxxdx, obliquantur; quarum ordinem duntaxet simplicem per Musculi corpus ductum, aliqui obtinent. Aliis autem, duplex series contigit; quæ utrinq; ex tendine quodam recto qui ab initio ad finem Musculi pergit, enascuntur. Credibile est sanè cuiq; ferè Musculo peculiare artificium inesse, ut in Cordis potissimùm fabrica (quod Hippocrati μῦς ὃ τι κατα τι ἰηρὰ) luculenter apparet: ut tum ad motum præstandum, tum ad figuram decentem toti corpori conciliandum foret opportunior: Atq; optârim herclè, ut alii quibus ad id otium est, hanc rem curiosiùs attingant. Haud dubitem enim, totam Musculi contractionem ac motum, à quibuscunq; causis accidat, hinc multùm adjuvari.

22. Tertiò, Olim proditum est à summo isto Anatomico Vesalio in aliquibus Musculis fibras hasce plures esse ac magis à se dissitas quàm in aliis: Quinetiam in diversis ejusdem Musculi locis, aliter ac aliter ab invicem distare. Quantum ego oculis usurpare possum, semper propè istam Musculi partem, quæ in ipsius contractione tumescit, spatia sive intervalla inter singulas fibras laxiora sunt; propè utrûmiq; autem extremum (quod Musculi fabrica omninò requirebat) constringuntur. Quod etiam magis confirmat summus Vesalius qui ventrem Musculi illic propriè reponen-

21. Secondly, the fibers with which we deal not only have different directions depending on the position of the parts to be moved, but also (and this has not been adequately observed by anybody so far) the fibers of the same muscle destined to the same movements are completely entwined in various entanglements and are intermingled with each other, as can be seen in the semitendinosus muscle of the human body which you would think is made as if from other muscles enclosed within it. In some [muscles] the fibers proceed directly from the head to the tail. In others they are twisted more or less along spiral or helicoidal lines. Some of these maintain just a simple arrangement through the body of the muscle. In others, however, there are two series; these originate from either side of a certain straight tendon which runs from the beginning to the end of the muscle. One can reasonably believe that there is a particular arrangement for almost every muscle, as appears clearly above all in the structure of the heart (for Hippocrates the heart is a strong muscle) so that the arrangement will be more suited either to carry out a movement or to provide the whole body with a seemly shape. Assuredly I would have preferred that others who have the leisure for it, would show more interest to tackle this matter. Indeed I do not doubt that all the contraction and movement of a muscle, whatever the cause, would be much helped by this.

22. Thirdly, the great anatomist Vesalius has shown long ago that these fibers are more numerous and more separated from each other in some muscles than in others. Even in different parts of the same muscle they are at different distances from each other. As far as I can perceive with my eye, it is always near that part of the muscle which swells when the muscle contracts that the spaces or intervals between different fibers are wider. But near any extremity they are drawn together (which is generally demanded by the structure of a muscle). The great Vesalius confirms this. He says that the place where the fibers of the muscle are drawn less closely together

15

ponendam esse (ait,) ubi Fibræ Musculi invicem minus
compinguntur ac copiosiore carne continentur.

23. Quod autem superius de Nervis monui, simile
de spinali Medulla accipiendum est, quam si in aquá per
dies aliquot maceres, apparet ipsam ex infinitis quibq; funiculis ac filis, Caudæ instar Equinæ, constare: ac Probabile est, non aliam esse Cerebri (ut supponunt quidam Cartesiani) texturam.

24. Post jacta hæc fundamenta, ad Hypothesin nostram
de ratione motús Musculorum explicandam aggredior: nihil enim in re obscurissimá fidenter affirmo, sed nudam
duntaxat Hypothesin proponere institui, idq; breviter; ac
proinde non pauca, quibus hæc omnia fusius probari possent
consultò omisi: namq; eam vel rejicere vel ulterius posthac
excolere animus est, prout futuris vel meis vel aliorum experimentis, magis minúsve congruere animadvertam. Esto
in Fig. 1. Musculus aliquis ante contractionem A B C D,
cujus fibræ ab A rectà porriguntur ad Tendinem C; suppono autem juxta ea quæ Sect. 12. & in fine Sect. 13. & Sect.
16. antè tradita sunt, harum fibrarum singulas, ut fibram
x T C, & singula quæq; intervalla sive spatia inter hasce
fibras, ut o a C, A P C quæ carne ubique repleri & in-
crustari diximus, Non tantum latice quodam spirituoso sui
generis turgere, unde nutriuntur & incrementum capi-
unt; verùm etiam hosce spiritus, tum intrà singulas fibras, tum quoque intrà singula ista spatia ordinatim
quidem ac continuá serie ità dispositos esse, ut, accedente aliquo impulsu, omnes intrà proprios suos cancel-
los, id est, intrà poros ductúsque sibi per totam Musculi longitudinem excavatos, per rectas lineas ab A ad
C, aut à C versùs A, ferantur: aut alius peculiari quodam modo juxtà Sect. 21. incurvatas. Neq; posse ipsos,
dum

and are enveloped in more flesh must be properly considered as the belly of the muscle.

23. What I said above about the nerves holds for the spinal cord as well. If you steep it in water for some days, the cord itself also appears to consist of an infinity of tiny strings, like a horse tail. And, as some disciples of Descartes assume, the texture of the brain is probably not different.

24. After having laid down these foundations, I proceed to explain our hypothesis on the reason of the movement of muscles. I do not indeed boldly assert anything on [this] very obscure matter, but I have resolved to put forward a mere hypothesis, and that briefly. Therefore, I have deliberately omitted many points by which all these matters could be proved more amply and, indeed, I intend either to reject the hypothesis or to develop it further inasmuch as I observe that it is more or less in agreement with future experiments either by myself or by others. In Figure 1 let ABCD be a muscle before its contraction. The fibers from A extend straight to the tendon C. Together with what was said previously in Section 11 and at the end of Section 13 and Section 16, I suppose all separate fibers such as the fibre xTC, and the separate intervals or spaces between these fibers, such as oaC, APC which we have said are everywhere filled full and encrusted with flesh, are not only swollen by a certain spirituous fluid specific to them from which they are fed and grow, but also that these spirits are distributed within the individual fibers as well as within those individual spaces regularly and in a continuous series so that, if some impulse occurs, all of them within their own enclosures, i.e. inside the pores and ducts hollowed through the entire length of the muscle, are carried in straight lines from A to C or from C to A, or other lines which are bent in some particular way according to Sect.21. And [I suppose that],

24

dum omnia fe rectè in corpore habeat, extrà hæc fpatia
evsgari. Hoc autem non iniquè à me peti, ex primo illo
Sect onis 18. Experimento fatis elucefcit. Poftquam hæc
fcripfiffem, incidi in locum Cl. Fallopii in Obf. Anxtom.
ubi hoc pluribus argumentis in Cl. V. fufiùs probatur, è
quibus pauca hic propter fummam viri autoritatem ex-
fcribam. ' Exemplo funt Mufculi (inquit de partibus agens
' quæ motu moventur voluntario) qui feipfos contrahunt
' & extendunt : cùm autem in unaquaq; particulâ motus
' certus fit & determinatus ad aliquam pofitionem, neceffe
' erat, ut corporis fubftantia ad iftam pofitionem effet di-
' recta, nè incerto motu agitaretur: ea directio eft coordi-
' natio minimarum particularum ad ipfam pofitionem di-
' ctam; hinc duo fequuntur, ut dentur meatus ordinati ad
' illam pofitionem, & ut fibras habeat illud corpus : Fibræ
' autem nihil aliud funt quàm minimæ illius fubftantiæ ali-
' quâ ferie ac ordine compactæ. Cúmq; motus voluntariè
' agens fit fpiritus, hic neceffariò requirit meatus in quibus
' fedeat; indéq; neceffe eft etiam ut fpiritus ifti contenti &
' difpofiti fint ad eandem pofitionem motûs: ac adeò ordi-
' natam locationem requirunt. Hinc patet, pro ipfis ordi-
' natos requiri meatus; at ordinati meatus in aliquo corpore
' neceffariò efficiunt fibras, &c.

25. Cæterum ut pergam ad reliqua; Sit S E pars trunci
medullæ fpinalis è cerebro prodeuntis, & ex infinitis, ut
dixi, fibris conflata; ex quâ ad E pronafcatur Nervus aliquis grandior E F G, qui circà G in varias propagines
finditur; earum quæ Mufculum A B C D perfodit, in ip-
fumq; eo fe modo quo Sect. 11. dictum eft diffunctit, efto
q r N, cujufq; flexus membranofi Intelligantur effe pun-
ctuli 3. r m. k. n 4 8. r. aliq; intra mufculi habitum in-
numeri. Sumantur jam duæ lineæ, H q r N. S q P y.

quartum

while everything functions properly in the body, they cannot get out of these spaces. Actually that first experiment of Section 18 sufficiently demonstrates that I am not wrong in giving this explanation. After having written this, I happened upon a passage in the *Obs. Anatom.* of the very famous Fallopius in which the famous man proves this at great length with many arguments, a few of which I will quote here because of the very great authority of the author. "There are for example muscles which contract and extend themselves (he says when dealing with the parts of the body which are moved in a voluntary movement). Since, however, movement is established and determined in each separate particle for a certain position, it was necessary that the substance of the body be directed for that position and be not agitated in uncertain movement. This direction is the coordination of the smallest particles to just that said position. This has two consequences: that there are channels arranged for that position and that this body has fibers. The fibers, however, are nothing other than the smallest [parts] of this substance compacted in a certain sequence and order. Since the voluntary agent of movement is spirit, this necessarily requires channels in which to reside. Therefore, it is also necessary that these spirits be contained and arranged in view of this position of movement. They thus require a well ordered location. From this it is obvious that ordered channels are required for the spirits. But, in a body, ordered channels necessarily produce fibers, etc."

25. Now to proceed to what remains [to be said], let SE be a part of the trunk of the spinal cord coming from the brain and made up, as I have said, of an infinity of fibers out of which, at E, there originates some large nerve EFG. Near G it divides into various branches, one of which pierces the muscle ABCD. It spreads out into the muscle in the way described in Sect.11. Let it be qrN, of which the membranous offshoots are to be taken to be the little points, 3, z, m, h, n, 4, 8, t and countless others within the coating of the muscle. Let us take two lines HqrN and SqPy.

quarum prior à puncto Cerebri H, per Medullam spinalem
ac Nervum E F G per ramum ejusdem q r N, usq; ad N
&c. continuatur.   Posterior autem à puncto S circa punctum q ab illa deflectens, in alium ejusdem Nervi ramum
q P y usq; ad y producitur : Suppono autem eas sic intrà
spinalis Medullæ ac Nervi E F G membranas incedere, ut
se nullibi contingant. Porrò utramq; harum linearum fibras
minutissimas infinitas, eodem modo extensas repræsentare
volo. Concipiendum est autem eandem planè rationem
esse reliquarum fibrarum, quæ in totâ Medullâ spinali &
omnibus corporis Nervis existunt.

Quas inter fibras, substantiam quandam medullarem ubiq;
impactam esse, fatentur quidem Anatomici, & res ipsa
ostendit. Illa verò quæ superiori Sectione de Spirituum
per Tendinosas Musculi fibras ordinatione ac τἔη dicta
sunt, pariter de Spiritibus qui in fibrillis hisce, ac etiam
spinalis medullæ Nervorumq; Tunicis, & medullâ spon-
giosâ insunt, proportione intelligi debent. Tandem esto
Arteria quæ ad Musculum eundem procurrit, & intra
ipsum divaricatur I K O, vena socia sanguini revehendo
dicata L M N.

26. His positis, existimemus animantis Corpus nihil
aliud nisi Machinam quandam aut ἀυτόματόν esse ; Animám-
que, quæ in nobis est, ab ipso tantisper cogitatione amo-
veamus, vel saltem ut spectatorem tantùm fabulæ hujus
quæ in corporis scenâ peragitur, in cerebro assidere singa-
mus. Certissimum est, ex solâ spirituum fluctuatione quæ
in cerebri medullâ soat, ac caloris vi continuò agitantur,
Musculos quoslibet hujus machinæ absq; ullâ omninò animæ ope aut concursu Mechanicè moveri posse : quod è
morbis convulsivis Sect. 8. suprà satis probatum fuit.
Et admodum verisimile est, omnes Infantium recens edito-

rum

The first extends from a point H in the brain through the spinal cord and the nerve EFG, through the branch qrN of the latter down to N, etc. The other one originates from a point S, is deflected by the former about point q into another ramification qPy of the same nerve and proceeds down to y. But I suppose that they run inside the membranes of the spinal cord and of the nerve EFG in such a way that they nowhere touch each other. Further, I intend that each of these lines represents an infinity of very tiny fibers extending in the same way. It should, moreover, be understood that the same explanation applies to the other fibers which are present in the whole spinal cord and in all the nerves of the body.

Also the anatomists say that a certain medullary substance is impacted everywhere between these fibers, and so it appears. Indeed, what was said actually in the previous section about the ordered distribution of the spirits through the tendinous fibers of the muscle should be understood as applying likewise to the spirits which are present in these fibrils, as well as in the tunics of the spinal cord and of the nerves, and in the spongy medulla. Finally, let the artery which proceeds to the muscle and spreads out inside be called IKO, and the associated vein for returning the blood be called LMN.

26. This being laid down, let us consider the body of a living creature as being nothing else than a sort of machine or an automaton and let us, for a while, banish the mind which is in us or at least let us imagine that the mind is seated in the brain as a mere spectator of this play which is taking place on the stage of our body. It is quite certain that from the flow of the spirits which are in the medulla of the brain, and continuously agitated by the force of heat, any muscles of this machine can be moved mechanically without any influence or contribution at all of the mind. This was sufficiently proved from the convulsive movements mentioned in Sect.8. And it is very likely that all the movements of new-born infants arise entirely

16

rum motus, penitus ab eadem causâ provenire. Illa autem
quæ postmodùm accedit concinna ac decens membrorum
corporis Inflexio, Usu ac Institutione comparatur. Quod
ex historiâ Pueruli sylvestris in Lithuaniâ nuper capti (ut
alia taceam) manifestum sit: qui Ll finis plane moribus
(apud quos degebat) ac incessu fuit. Notissimum quoq;
est, Infantes pedum usum nescire, donec teneelli eorum
gressus, à parente aut nutriculâ conformentur. Quòd non
eo dico, quasi motus isti quos usu ac aliorum imitatione
didicimus, minùs essent Mechanici; aut, accedente aliquâ
actione animæ, non per solos motus corporeos, purè ac
merè efficerentur; sed quod Machina hæc nostra, aliter in
motibus ordinatis ac regularibus, aliter verò in Istis quos
dixi inconditis afficiatur. Quæ sanè ut ritè intelligi que-
ant, notandum est, Sensum (quod Cartesiani facilè concedunt) esse, Actum animæ quo motum quendam in particulis sive fibrillis Cerebri excitatum percipit: quomodo
autem motus in ipso fieri contingat, suprà Sect. 17. expli-
catum est. Quibus jam addo: Cùm omnis sensatio in
membranâ aliquâ fiat, omnes verò membranæ sensoriæ à
cerebri meningibus oriantur, eóq; modo se habeant quo
ibi dictum est; necessariò illud consequitur, ut percussio
sive motus in membrana organi sensûs ab objecto factus,
tandem ad cerebri Membranas perducatur. Sciendum
porrò est, eas infinitas ex se fibras in totam cerebri substantiam quaquaversùm emittere; cúmq; à diversis sensuum organis aliæ ac aliæ producantur, etiam intrà cerebrum separatæ ab invicem ac distinctæ sunt, ut de fibris
spinalis medullæ ac Nervorum monuimus: Et quam rationem fuisse diximus, Spirituum intra fibras Musculorum ac
Nervorum, eadem planè & hîc est: Quia verò ipsæ membranæ pulsari nequeant, quin fibrillæ hæ simul moveantur;

Hinc

from the same cause. But that skillful and seemly bending of the limbs of the body, which occurs later on, results from practice and training. This appears clearly from the story of the little forest boy recently captured in Lithuania (not to mention other examples). He had fully adopted the manners and the gait of bears (with whom he lived). It is also well known that infants do not know how to use their legs until their tender baby steps are fashioned by a parent or a nurse. I do not mean that these movements which we have learnt by practice and the imitation of other people are less mechanical or that, when some action of the mind occurs, they are not accomplished purely and simply through body movements alone. But I mean that this machine of ours is managed in one way in the case of ordered and regular movements and in another way in the case of those which I have called disordered. To make the matter more clearly understood, it must be noted that sensation (which the disciples of Descartes readily admit) is an act of the mind by which it perceives a certain movement excited in the particles or in the fibrils of the brain. It has been explained above in Sect.17 how movement can occur in the brain. To this I add: since all sensation is produced somewhere in a membrane and all sensitive membranes in fact originate from the meninges of the brain and behave in the same way as described above, as a consequence, percussion or movement made by an object in the membrane of an organ of a sense is finally transmitted to the membranes of the brain. It should be known, however, that these membranes send out an infinity of fibers everywhere into the entire substance of the brain. Since different fibers come from different organs of the senses, they are also separated from each other and distinct inside the brain, as we pointed out about the fibers of the spinal cord and of the nerves. And for the reason we gave, exactly the same arrangement of spirits within the fibers of the muscles and of the nerves, is also present here. Since, in fact, the membranes themselves cannot be set in motion without these fibrils being moved simultaneously,

17

Hinc à variis membranarum in organis sensuum, adeóq; harum fibrillarum cerebri ictibus, universæ rerum Ideæ in cerebro existunt, & ab animâ percipiuntur: Namq; (quod alias fusiùs ostendi) ea est Animæ in hac Machinâ conditio, ut omnis particularum motus, qui in organo ali-quo sive externi sive interni sensûs fit, ab ipsâ necessariò percipiatur.

27. Percipi dico; quia nonnulli præter hujusmodi per-ceptionem, nihil omninò aliud negotii Animæ concedunt; indéq; opinantur, omnia quæ in animantibus contingunt curvata, ex principiis purè Mechanicis utcunq; explicari posse: Equidem gravissimas hic difficultates se offerre ul-trò fatentur; at profectò (ut ipsis videtur) non minores in alterâ sententiâ futuræ sunt, quæ vim quandam spiritus animales huc & illuc, in illum vel istum vel alium Nervum aut Musculum, pro arbitrio impellendi attribuit: Namq; exempli gratiâ; cùm adest brachii aut cruris movendi vo-luntas, Unde novit anima,qui Nervi Musculive in hac cor-poris fabricatione, ad istum usum destinantur? Quicquid sit, sive hæc sive altera hypothesis pro verâ habeatur; neutra tamen, illa quæ de modo sentiendi dicta sunt labe-factabit: utramq; igitur ad motum Musculorum expli-candum adhibere possumus. Redeamus ergò jam ad Fig. 1. & concipiamus à quacunq; sit causâ, adesse animo volun-tatem movendi Musculum A B C D. partémq; ipsi ad-nexam. Dico primùm, in hoc casu, istas tantum fibrillas quas repræsentare diximus lineam H q r N impulsu aliquo undecunq; sit sive ab ipsâ Animâ volente, sive à causâ ma-teriali, in extremitate superiori H, agitari, immotâ pror-sùs alterâ S q p y, & reliquis universis. Hinc enim fit, ut vis movens quam supra ex cerebro per Nervos exire proba-bamus, perveniat ad ramum G r N, indéq; ad Musculum

D   A B D

then from various strikings of the membranes in the sense organs and thence of the fibrils of the brain, general concepts of things are present in the brain and are perceived by the mind. And indeed (as I have shown more fully elsewhere) the condition of the mind in this machine is such that all movement of particles which takes place in any sense organ, either external or internal, is necessarily perceived by the mind.

27. I say "is perceived" because several people concede absolutely no other occupation to the mind beyond perception of this sort. They claim that all phenomena which happen in living creatures can be explained one way or another by purely mechanical principles. Certainly they do further admit that the very gravest difficulties present themselves on this point, as in fact (as appears to these same people) there are likely to be no lesser [difficulties] in the other opinion which ascribes some force to impel the animal spirits hither and thither in this, that or the other nerve or muscle at will. For example, if one wishes to move an arm or a leg, from where does the mind know which nerves or muscles are destined for that purpose in this structure of the body? Anyway whether this or that hypothesis is held as true, neither will upset what has been said about the manner of feeling. We can consequently apply one or the other to explain the movement of the muscles. Let us then return now to Fig.1 and let us imagine that, for whatever reason, the desire to move the muscle ABCD and the limb attached to it is present in the mind. I claim firstly, in this instance, that only those fibrils alone which we said the line HqrN represents are shaken at their upper extremity H by some impulse whatever its origin, either by the mind willing it or from a material cause, whereas the other [line] Sqpy and all the rest are immobile. Hence it indeed happens that the moving force which we proved above to go out of the brain through the nerves, arrives to the branch GrN and from there to the muscle

18

A B C D cui is inferitur, non autem ad Mufculos illos, ad quos pertinent rami S q P y, S q P Z. Et eandem planè rationem in cæteris omnibus corporis Mufculis movendis aut non movendis, (pectare debemus. Quòd autem certarum partium motus à certis Medullæ fpinalis Nervo únꝗ; fibris pendeat, primò à texturâ Nervorum ac medullæ fpinalis, de quâ Sect.23. dictum eft, fuaderi poteft; deinde vel unicâ hac inftantiâ haud parùm confirmatur: Vidi (inquit Laurent.) Adolefcentem Nobilem, qui Vulnere in fpinali medullâ accepto, ftatim cruris ac pedis dextri motu privatus eft, fuperftite utriúfq; brachii & totius ferè corporis motu: proptereà quòd, illæ fibræ (ut ipfe quoque Laurentius difterit) quæ ad has partes pertigerunt, læfæ ac incifæ fuerint; aliæ autem minimé. Hinc etiam quatuor digiti diftenduntur, variiſq; motibus flectuntur, ab unico extenfore Mufculo, in quem unicus tantum Nervorum Surculus (ut annotat. Highmorus) implantatur. Cæterùm, penes Animam effe (ut ex iftâ hypothefi philofophemur) Fibrillas folas lineæ H q r N, aliis non agitatis, movere; haud certè magis mirandum eft, quàm poffe eam in Nervi ramulum q r N ita fpiritus tranfmittere, ut folus Mufculus A B C D moveatur, cæteris ad quos pertingunt ramificationes q P y, q P z quiefcentibus; id quod omnes fatentur: Aut ex infinitis, quæ in memoriâ funt, rerum Ideis, quamcunque velit ad contemplandum pro lubitu feligere: Illæ enim Ideæ funt tot diverfi motus (ut maximè probabile eft) in diverfis totidem cerebri particulis feu fibrillis confervati: Certè, nifi materiales effent, non poffent à morbo deleri, ut in his quos omnium rerum occupavit oblivio. Si verò ex alterâ hypothefi loquamur, animadvertendum eft, Omnes iftas Ideas, diftinctis quidem locis, & certo quamq; ordine in cerebro collocari: quòd hi faciliùs conceſſerint, qui

nórunt

ABCD into which it is inserted, but not to those muscles to which the branches SpPy, SpPz belong. And we ought to consider exactly the same explanation for the movement or non-movement of all the other muscles of the body. The movement of certain parts depends on certain fibers of the spinal cord and of the nerves. This can be supported firstly from the texture of the nerves and of the spinal cord as described in Sect.23. Then the following instance, although unique, confirms this very well. "I saw," Laurentius says, "a youth from the nobility who, after his spinal cord had been injured, was deprived forthwith of the movement of his right leg and foot whereas he kept the movement of both arms and almost all his body." [This occurred] because the fibers which go to these parts (as Laurentius himself explains) were injured and cut-through, whereas the others were not. Hence also, four fingers are extended and flexed in various movements by a single extensor muscle alone in which only one branch of the nerves is implanted (as observed by Highmore). For the rest, supposing that it is within the power of the mind (if we reason from that hypothesis) to move only the fibrils HqrN while the others remain immobile, it is certainly not to be wondered at that the mind is able to transmit spirits through the small branch of the nerve qrN in such a way that only the muscle ABCD moves whereas the others supplied by the branches qPy, qPz remain at rest, a thing that everybody admits; or that from the infinity of concepts of things which are in the memory, [the mind] chooses any one it would wish to contemplate according to its liking. These concepts indeed are as many different movements (as is very likely) preserved in just as many different particles or fibrils of the brain. Certainly, if they were not material, they could not be destroyed by a disease as occurs in those seized by oblivion of all things. If in fact we speak based on the other hypothesis, it must be observed that all these concepts are located in certain distinct places and in some certain order in the brain. Those who have come to know how distinctly and

autem quàm diſtinctè ſeparatéq; infaits propemodùm
Horizontis objecta in parta oculi camera depingantur, deinde ex Mania conſtat, quæ iſtius ordinis perturbatio quædam eſt, idearúnq; confuſio. Notandum etiam hunc ordinem non alium eſſe, quàm qui à diverſis temporibus
ſumitur, in quibus, una poſt aliam in cerebrum pervenit:
& hinc etiam diverſas in eo ſedes obtinuére. Illud autem
hinc apertè conſtat quòd res aliqua fortè conſpecta, aliam
à ſe & natura & tempore long ſſimè alienam, cujus utpote
ante aliquot annos ne minima quidem inciderat cogitatio,
ſæp ſſimè nobis in memoriam revocet. Inde procaldubio,
quod ambarum ideæ diu anteà & loco & tempore con-
junctæ fuiſſent; & cùm jamdudùm in cerebro ſimul la-
tuiſſent, alterâ per hanc occaſionem evocatá, non potuit
altera non apparere: hinc etiam Memoria (uti vocatur)
Localis tota pendet. His præmiſſis, exiſtimo dici poſſe, Ab
impulſu alicujus objecti externi vel interni, primò in fibrillis cerebri talem motum excitari, qui ab animâ per-
ceptus in ipsâ neceſſariam movendi cruris voluntatem
efficit, ac ſimul (nihil prorsùs ad hoc conferente iſtâ vo-
luntate, ſed cum illo impulſu neceſſariò ſemper coexiſten-
te) ex Mechanicâ partium conſtructione ad iſtum cerebri
locum præcisè defertur, in quo ſunt aut vocum Ideæ quæ
crus aut brachium movendum ſignificant, aut Ideæ cruris
ac brachii moti, aut ambæ ſimul: Ideæ autem iſtæ ab
primâ uſq; infantiâ eum in 'cerebro locum occuparânt, ubi
erant extremitates fibrarum lineæ H q r N aut S q P y,
quia cùm primùm iſta membra movere ex aliis didicimus,
fuimus ad hoc faciendum ſermone admoniti, aut conſpe-
ctu ac imitatione eorundem in aliis motuum inſtituti.
Idea namq; cujúſq; rei faciendæ ex neceſſariâ Fabricæ hujus corporis ratione, ad eam cerebri partem devenire cen-

D 2   ſenda

separately the almost infinite number of objects at the horizon are depicted in the small chamber of the eye will admit that more easily. It appears also in madness, which is some disturbance of this order and a confusion of the ideas. It should be noted that this order is nothing else than the one which results from the different moments at which the ideas arrive in the brain one after the other. Hence, they have occupied different places in the brain. This moreover is clearly established from the fact that something which happened to be observed by chance, often recalls in our memory another thing far different from the former in nature and in time, the thought of which has never occurred in the memory for some years. Thus, unquestionably, the concepts of both things had been long ago linked in place and in time. Since they had been hidden in the brain together for some time, when one [idea] has been evoked on this occasion, the other could not be prevented from appearing. All local memory (as it is called) depends on this. After these premises, I think it can be said that it is by the impulse of some object, external or internal, that movement is first aroused in the fibrils of the brain such that having been perceived by the mind, it produces in it the will necessary to the moving of a leg and simultaneously (that will contributing nothing directly to this but necessarily coexisting with that impulse), as a result of the mechanical construction of the parts, it is transmitted precisely to that part of the brain in which there are either the concepts which mean that a leg or an arm must move, or the concepts of a leg or arm having been moved, or both together. Moreover these concepts had, from early childhood, taken possession of that place in the brain where the extremities of the fibers of the line HqrN or SqPy were, since when first we learnt from others to move those limbs, we were reminded to do that by words or instructed by seeing and imitating these movements in other people. The idea indeed of doing something should, from the required systematic structure of this body, be thought of as coming to that part of the brain in which there are

20

fenda eft, in quâ funt extremitates fibrarum ad id requifi æ.
Quo! nè dutiùs fonet, confiderare nos jubent paulifper,
duo illa è multis quæ in corpore noftro fæpiùs contingunt ;
Sternutationem dico, & Perpetuum iftum deglutiendi
conatum, quæ in uvulæ procidentiâ adeft : quorum alterum (ab Ariftot. parvâ Epilepfia vocatum, quodq; magno
plurium Mufculorum nixu peragitur) plumulâ quandóq;
aut leviffimo ftramine narium membranam tangente, excitari videmus : Alterum verò ab uvulæ extremitate membranam Oefophagi pariter attingente accidit. Ex his apertè conftat, itâ comparatam effe corporis ftructuram, ut
quoties membrana narium fic, uti diximus, afficitur, Motus
quidam ad cerebrum delatus iftas duntaxat fibrillas exagi-
tet, quæ diaphragma cæteróſq; quæ in fternutatione moventur, Mufculos ad agendum proritet. Confimiliter cùm
membrana Oefophagi, eos tantummodo qui deglutitio-
nem efficiunt. Quod certè Mechanicè omninò præftari,
five eo quo Sect. 26. innuimus, five alio aliquo modo fiat,
eft manifeftiffimum : Námq; vel affirmare oportet, Animam fuâ fponte, animales fpiritus in nervos Mufculofq;
fternutationi aut deglutitioni fervientes propellere ; atqui
hoc verum non eft, quia hæc vel maximè contra voluntatem noftram contingunt; deinde, Quæ obfecro eft ifta ſcien-
tia, quâ anima ex perceptione affectæ narium membranæ
fpiritus in Mufculos fternutatorios, Oefophagi autem in
deglutientes mittere novit ? Sin dicamus hoc ab animâ neceffariò fieri, perinde eft ac fi id Mechanicè tantùm abfq;
ullâ ejus ope accidere dixeris. Hæc fortaſſè aut his fimilia
ad hanc hypothefin excogitari poffent : Utcunq; fit, inquam
Impulfu illo quem accipit hujus nervi extremitas H in cerebro, non modò quaffatur tota illa fibrarum feries quæ per
lineam H qt N repræſentat: fuppono, ab ifto puncto H

the extremities of the fibers required for it. So that it does not sound too difficult, let us for a while consider two phenomena among many which often occur in our body. I mean sneezing and that continuous attempt at swallowing which is present in a prolapse of the uvula. We see that the former (called small epilepsy by Aristotle and which is carried out with the strenuous effort of several muscles) is caused sometimes when a feather or a very light straw touches the membrane of the nostrils. The latter similarly occurs when the extremity of the uvula touches the membrane of the oesophagus. These facts obviously show that the structure of the body is so arranged that whenever the membrane of the nostrils is thus touched, as we have said, some movement having been transmitted to the brain stirs up precisely those fibrils which stimulate into action the diaphragm and the other muscles which move during sneezing. Similarly, with the membrane of the oesophagus, they stimulate only those muscles which carry out swallowing. It is quite obvious that the process is completely mechanical, either in the way which we indicated in Sect.26 or in some other way. Indeed, alternatively we would have to assert that the mind spontaneously propels the animal spirits into the nerves and muscles which are used in sneezing and swallowing. But this is not true, since these phenomena occur very much against our will. Further, I ask, what is this knowledge by which the mind knows, from the perception of the affected membrane of the nostrils, to send spirits into the muscles involved in sneezing or [from the perception of the affected membrane] of the oesophagus into the muscles involved in swallowing? But if we say that this is necessarily done by the mind, it is just the same as if you were to say that this occurs only mechanically without any need of the mind. Perhaps these explanations or others similar to them could be devised for this hypothesis. However that may be, I say that by that impulse, which that extremity H of this nerve receives in the brain, not only is that whole series of fibers which I suppose is represented by the line HqrN, shaken from that point H

usq; ad extremos istos membraneos flexus in quos nervum
intrà Musculum definere dixi, verùm etiam per istim
concussionem, guttulas admodum multas ex punctis 3. 7.
m. k. n. 4 8. aliisq; propè innumeris, generofi istius quem
dixi liquoris eodem momento, exprimi, & in totum Musculi
habitum extillare opinor. Cùm enim jam satis probatum
sit, vim quandam è cerebro per nervos advehi in Musculum, nec, si oculis fides habenda sit, quicquam in nervis
appareat, quod huic usui magis convenire queat, quàm
opulentissimus ac spirituosus iste succus, qui constanti
circuitu per omnes nervos traducitur; Quid, obsecro, magis
verisimile est, quàm vim illam cum hoc liquore deferri,
aut potiùs esse hunc ipsum liquorem sive spiritum anima-
lem fibrarum impetu è Nervorum ramulis excussum?
Quod si sit, illud quoq; admodùm probabile erit, Ex ad-
mistione liquoris hujusce sive spiritûs cùm spiritibus sanguinis, continuò spirituosarum omnium particularum,
quæ in vitali totius Musculi succo insunt, magnam agitationem contingere, uti cùm spiritus vini spiritui sanguinis
humani admiscetur. Namq; omnem animantis partem vivifico quodam ac spirituoso liquore turgescere, suprà quidem
monui, ac omnibus est in confesso; ac nemo ferè tam in
Chymiâ hospes est, qui nesciat, Quanta particularum com-
motio ac agitatio, ex variis inter se permistis liquoribus accidere soleat. Ut in Exemplo modò allato, ac etiam in aquâ
communi oleo vitrioli, aut Butyro Antimonii spiritui Ni-
tri:ffuso, aliisq; tere hujus generis infinitis, cernere licet:
Hic autem simile aliquid etiam evenire, neq; hæc temerè
confingi, Primò docet, somma illa in sanguine ad fermente-
scendum aptitudo ut è cordis motu clarè apparet: qui beneficio fermenti cujusdam prout à Cartesio, Hogelando, Regio
explicatur; aut flammæ vitalis (uti vult Clariss. Entius)
                                                        uneuneni

to these furthest membranous folds into which I have said that the nerve ends within the muscle, but also, by way of this shaking, I think that very many small droplets of that very rich liquor which I mentioned are expressed at the same moment from the points 3.z.m.k.n.4.8. and from almost countless others, and they are instilled into all the muscle. Since indeed it has now been proved sufficiently that a certain force is carried from the brain through the nerves into the muscle, and that, if the eyes can be relied on, nothing else appears in the nerves, which is more suited to this purpose, than that very rich and spirituous juice which is drawn through all the nerves in a constant circuit, what, I ask, is more likely than that the force is carried with this liquor, or rather that this liquor itself, or animal spirit, is driven out of the branches of the nerves by the violent movement of the fibers? If this be so, it will also be highly probable that from the mixture of this liquor or spirit with the spirits of the blood, a great agitation of all the spirituous particles which are present in the vital juice of the whole muscle occurs continuously, as when spirit of wine is mixed with the spirit of human blood. For I have stated above that every part of a living creature is swollen by a certain vivifying and spirituous liquor and this is acknowledged by all. No one is such a novice in chemistry as not to know how great a commotion and agitation of the particles usually occurs from different liquors mixed with each other. This can be seen in the example just mentioned and also in plain water mixed with oil of vitriol, or in butter of antimony poured into spirit of nitre, and in an almost infinite number [of other instances] of this kind.[4] But I have not imagined without reason that here something similar also happens. Firstly, this is suggested by that extreme ability of the blood to ferment - as appears clearly from the movement of the heart - which is accomplished by means of a sort of ferment, according to the explanation of Descartes, Hogeland, and Regius, or by means of a vital flame (according to the very famous Ent)

---

[4]See Introduction, p. 30

22

sanguinem perpetuò rarefacientis perficitur. Ostendit præterea vis illa, quam in totâ sanguinis massâ fermentandâ obtinet materia seminalis, ut in His quiталlientibus, castratis, ac Oestro venereo percitis, observare possumus. Præcipuè verò ex affectibus hystericis, ubi sanguis, corrupti seminis miasmate inquinatus, miram fermentationem concipit, unde in Pulmonum vasis (ut explicat Eruditiss. Highmorus) turgescens, suffocationis periculum minatur; quinimo omnem Nervorum succum eadem labe infici probabile est. Cæterùm materia hæc seminalis ex opt. Anatomicorum sententiâ, nihil aliud est præterquàm redundantia quædam laticis Nervosi, suprà quam quæ in vesiculis seminariis glandulísq; vicinis ad futuram prolem colligitur. Siquis miretur, è tantilli liquoris aspergine tantam in succis istis quibus repletur Musculus tumultuationem cieri; meminerit, Illum unico momento ex infinitis propemodum nervi intrà Musculum propaginibus ad puncta 3. z. m. &c. egredi, & hinc luctam illam & conflictationem in pluribus partibus Musculi simul vehementer accendi. Quinetiam virtutem maximam, in exili admodùm materiâ delitescere, ex unico (si alia non suppeterent) croco Metallorum addiscimus; nam post aliquot annorum infusiones, unde vomitus sæpè immanes ac purgationes excitatæ sunt, vix quicquam ipsi ponderis decedit.

28. Porrò, cùm antè ostensum fuerit, Omnium qui in partibus Musculi sunt spirituum motum intra certa quædam spatia contineri, ac præterea spatia hæc (ubi venæ est Musculi), ampliora ac patentiora esse; consideremus, unum illud spatium A P C: Inquam, agitationem illam spirituum quæ fit à diversis liquoribus intra Musculi membranas tumultuantibus, necessariò ipsos impellere magno
nisu

which continuously rarefies the blood. Moreover, it is shown by that force which seminal matter exerts in the fermenting of the whole blood, as we can observe in pubescent boys (with the break of voice), in castrates, as well as in those excited by sexual frenzy, but above all by the hysterical states in which the blood polluted by the miasmata of corrupted semen conceives an astonishing fermentation so that it swells in the pulmonary vessels (as the most learned Highmore explains) and danger of suffocation threatens.[5] Moreover it is likely that all the juice of the nerves is tainted by the same disease. For the rest, this seminal matter, according to the opinion of the best anatomists, is nothing other than a certain excess of the nervous fluid in addition to that which is collected in the seminal ducts and adjacent small glands for the future offspring. If somebody wonders that such turbulence is stirred up in these juices with which the muscle is filled, by a sprinkling of so little liquor, he should remember that this liquor gets out instantly from an almost infinite number of offshoots of the nerve within the muscle to the points 3.z.m. etc, and hence that struggle and conflict is ignited simultaneously in several parts of the muscle. Even more, we learn from the unique crocus [of metals] (if other examples were not available) that the greatest virtue is concentrated in very little substance. For, after infusions over several years, whereby vomiting, often excessive, and purges, have been provoked, scarcely any of its weight has left it.[6]

28. Moreover, since it has been shown previously that the movement of all the spirits which are in the parts of a muscle is contained within certain spaces and, in addition, that these spaces (where the belly of the muscle is) are wider and more accessible, let us consider that single space APC. I say that that agitation of the spirits which results from the turbulence of the different liquors within the membranes of the muscle necessarily impels these liquors with a great effort

---

[5] See Introduction, Footnote 77, p. 30.

[6] See Introduction, Footnote 78, p. 31.

nifi per lineas rectas versùs A & C. Cùmq; intervalla
five spatia semper minora versùs extrema caduent, in
se reflectuntur, ac in Lateribus circà medium Musculi B,
five Ventrem (ubicunq; fit) in majori quantitate colliguntur: & hinc Musculus circa ventrem incipit Tumescere;
Quod idem de reliquis omnibus fibris pariter accipi
debet.

29. Dixi, hinc incipere Musculi extumescentiam, quòd
absq; aliâ quoq; causâ in susisum ascitâ, ex se quidem
hæc non satis ad ipsum movendum valeret: Quamobrem
cùm ferè omnes fateantur istiusmodi esse animæ in spiritus
animales imperium, ut ipsos ad varios membrorum motus præstandos quoquò velit transferre, & in quoslibet pro
lubitu nervos, per nudum voluntatis actum, propellere
possit, ut e.g. in nervum E F G N ad Musculum A B C D
movendum; quidni etiam arbitrari licebit, ipsam per
eundem volendi actum, sanguinem quoq; arteriosum, eodem momento, per arteriam I K O copiosiùs in eundem
Musculum protrudere? Certè qui in hac hypothesi sunt,
ultrò concedunt, Esse præterea animæ suum in totam sanguinis massam imperium, ut ex variis affectibus, Irâ, Lætitia, Pudore, Libidine, & aliis, abundè patet: in quibus
omnibus sanguis non modò in corde vehementer commovetur; sed etiam ad diversas corporis partes protruditur.
Et quia ex alterâ hypothesi de quà Sect. 27. in fine actum
est, dici potest, Materialem istam in cerebro Ideam, quæ animam in singulis passionibus afficit, eodem etiam instanti
sanguinem arteriosum in certas corporis partes mechanicè
impellere: Ideò secundùm hanc non absurdè quis dixerit;
Eandem Ideam, quæ in animo excitat Musculi A B C D
movendi voluntatem, ac spiritus simul animales ad illud
præstandum in nervum E F G N mechanicè deduxit, pariter
riter

along straight lines towards A and C. And since they strike [the sides of] intervals or spaces, always smaller towards their ends, they turn back on themselves and are collected in greater quantity in the looser spaces about the middle or belly of the muscle B (wherever it may be). And hence the muscle begins to swell about its belly. And the same has to be admitted about all the other fibers.

29. I said that the swelling of the muscle begins from this because this would not be sufficient by itself to move the muscle without invoking the assistance of another cause from without. Therefore, since almost everybody agrees that the control of the animal spirits by the mind is of such a kind that it can move these spirits wherever it wishes to carry out different movements of the limbs, and can propel them into any nerves it chooses through a single act of the will, as for example into the nerve EFGN to move the muscle ABCD, why should one not also believe that the mind, by means of the same act of will, at the same moment, thrusts forth more arterial blood through the artery IKO into the same muscle? In any case, those who hold this hypothesis admit willingly that the mind also exerts its control over the whole mass of the blood as appears clearly from different feelings: anger, joy, shame, lust and others. In all these feelings the blood is not only stirred up violently in the heart but it is also thrust out to the different parts of the body. According to the other hypothesis dealt with at the end of Sect.27, it can be said that that material concept in the brain which affects the mind in the different feelings likewise mechanically impels the arterial blood, at the same instant, into certain parts of the body. Therefore, according to this hypothesis, it would not be wrong to say that the same concept which excites in the mind the will to move the muscle ABCD and has at the same time mechanically driven off the animal spirits down into the nerve EFGN to carry out this movement,

24

riter eodem temporis momento intercedente nervo qui infinitis ramusculis in cordis auriculas inseritur, majorem è
corde sanguinis copiam per arteriam I K O in eundem propellere. Sed restat adhuc alia, non minùs mechanica, utq;
mihi videtur, valdè probabilis ratio sanguinis arteriosi citius ac copiosiùs per arteriam I K O in Musculum movendum A B C D advehendi.   Nam obortâ, quam dixi, in
toto Musculi habitu particularum tumultuatione ac effervescentia, primò istæ particulæ jam celeriùs agitatæ, majorem motûs gradum acquirunt, deinde ex hoc motu atq;
agitatione, à se invicem magis recedere, ac majorem locum occupare nituntur (ut in omni Fermentatione contingere videmus) hinc necessum est, ut omnes pori carnis
Musculosæ magis laxentur ac aperiantur ; adeóq; Musculus
A B C D tumescere incipiat ; ex quibus omninò consequitur, Sanguinem è corde semper cùm impetu quodam in
omnes corporis arterias simul expulsum, minorem resisten-
tiam offendere in arteriâ I K O, quàm in alia quacunque.
Primùm enim, ob majorem celeritatis gradum in particulis
omnibus Musculi A B C D acquisitum, istæ faciliùs sanguini arterioso locum cedunt, ac multò minori vi ab ipso
in Musculum exundante repelluntur : Accedit deindè, Ex
poris ita apertis ac dilatatis, longè ampliora spatia ad majorem sanguinis quantitatem excipiendam, intrà Musculum patere : unde liquet, fore ut sanguis & celeriùs & copiosiùs è Corde per arteriam istam I K O in Musculum
A B C D perveniat : Idem prorsus in Inflammationibus
contingit, itemq; in Doloribus, quos medici propterea tra-
here dicunt, ab effervescentia humorum in parte excitata.
Deniq;, quia caro circa ventrem cujusq; musculi (ubicunq;
sit) ut Sect. 23. è Vesalio diximus, copiosiùs incrustatur, id-
circo ibi potissimùm Musculus intumescit, idq; eodem
modo quo Sect. 28. declaratur.   30. Quòd

likewise and at the same moment in time, by way of the nerve which penetrates the auricles of the heart in an infinity of small branches, propels a larger quantity of blood from the heart through the artery IKO into the same muscle. But there remains as yet another no less mechanical, and as far as it seems to me, very likely explanation for a quicker and more abundant delivery of arterial blood through the artery IKO into the muscle ABCD which is to move. For once there has arisen throughout the whole muscle an agitation and effervescence of the particles, as I have mentioned, firstly these particles, now accelerated, acquire a higher degree of movement. Then, as a result of this movement and agitation, they strive to move further away from each other and to occupy more space (as we see happening in every sort of fermentation). Hence it is necessary that all the pores of the muscular flesh are more lax and open, and therefore, the muscle ABCD begins to swell. As a general consequence of these processes, the blood which has been expelled from the heart into all the arteries of the body together, always with the same force, meets less resistance in the artery IKO than in any other. Firstly indeed, because of the higher degree of velocity acquired by all the particles of the muscle ABCD, these particles yield to the arterial blood more easily and are repelled with much less force by this blood flowing into the muscle. Then, by reason of the pores having been thus stretched wide open, it happens that there are far more ample spaces available within the muscle to receive a larger quantity of blood. Consequently, the blood can arrive more quickly and more abundantly from the heart through the artery IKO into the muscle ABCD. Exactly the same thing happens in inflammations and in painful conditions which on that account the physicians ascribe to the effervescence of the humours excited in the parts. Finally, since there is more flesh about the belly of any muscle (wherever it may be), as we mentioned in Sect.22 according to Vesalius, it is here that the muscle swells most of all as we mentioned in Sect.28.

25

30. Quòd verò planè infignis fit arteriofi fanguinis o-
pera in Mufculis movendis, à quacunq; tandem caufa in
ipfos deferatur, ante omnia, ratio illa arterias venáfq; in
Mufculos implantandi vehementer admodum fuadet.
Quare ex ufu erit, id quod de Arteriis Veníiq; fuprà innui,
obfervare. 'Arteriæ (inquit Aquapendens) ac venæ velut
'rami per totum Mufculum divaricantur: & mox; Si ve-
'rum fit ex Ariftot: fenfum abfq; calore non effe meritò
'infignes arteriæ per Mufculi fubftantiam diffeminantur:
Imprimis autem confidare debemus eas, unà cum venis
in Mufculos inferi, eodem prorsùs ritu ac modo, quo
fuprà Sect. 10. de nervis eft dictum, quod præclarè obfer-
vavit Spigelius, cujus hæc funt verba. 'Eádem, inquit, ra-
'tione (de Nervorum infertione modò loquutus) Vena
'ac Arteria fpargitur. Semper enim aut principium aut
'medium ingreditur, finem nunquam, nifi in longiffimis,
'quibus à vicinis ramus tranfmittitur. Nunquam autem
'fuperficiales funt, fed femper (ut Clariff: idem Anatom:
'affirmat) ad interiora penetrant præfertim venæ, quibus
'etiam hoc peculiare eft, ut communi fuâ tunicâ, quam
'habent dum extra Mufculos incedunt, careant, ficq; fim-
'pliciffimæ fint. Quantum autem fanguinis & quàm facilè
in univerfos corporis Mufculos per arterias continuò ad-
vehatur, certe vel hinc manifeftum eft, quòd omnis Mufculi caro (quæ maximam ejus partem facit, & ex quâ totius corporis moles potiffinùm conftat) nihil aliud effe
videatur, nifi ifta fanguinis per fibrarum intervalla fluen-
tis portio, quæ earum frigiditate craffefcens intra ipfas
fiftitur, carnémq; Mufculofam conftituit: hoc verò ità
effe, ex iis qui fame enecantur apparet, in quibus Vafa fanguine quidem turgent, Mufculi autem penitùs collapfi fe: è
omni carne fpoliantur: namq; tantifper in hoc cafu fan-

L   guinis

30. The part played by the arterial blood in the movement of muscles is very important whatever the cause of being supplied to the muscles. This is strongly suggested above all by the way in which the arteries and veins are implanted into the muscles. Therefore, it will be profitable to observe what I hinted at above concerning the arteries and veins. "The arteries and the veins (Aquapendente says) spread out like branches through the whole muscle," and then: "If it is true, according to Aristotle, that there is no feeling without heat, it is right that significant arteries be disseminated through the substance of the muscle." First of all we must consider that they are implanted into the muscles together with the veins in exactly the same way which we described for the nerves above in Sect.10. This is what Spigelius observed particularly well. These are his words: "The vein and artery branch out in the same manner (referring to the insertion of the nerves). They always penetrate the beginning or the middle [of the muscle], never the end, except in very long muscles to which a branch is transmitted from the vicinity. However, they are never superficial but always penetrate to the inner parts (as the same famous anatomist affirms), especially the veins which have the peculiarity that they are without their common tunic which they have when they run outside the muscles. Thus they are very simple."[7] How much blood is carried continuously through the arteries into all the muscles of the body, and how easily, is indeed certainly apparent from the fact that all the flesh of a muscle (which makes up the greatest part of it and constitutes the bulk of the mass of the entire body) seems to be nothing else than that portion of blood flowing through the intervals between the fibers, which, thickened by their coldness, is brought to a halt among them and forms muscular flesh. This is apparent from those who die from hunger. In them the vessels are indeed swollen with blood but the muscles, having completely collapsed, are stripped of almost all flesh. In such instance,

---

[7] See Introduction, Footnote 79, p. 32.

26

guinis circulatio continuatur, donec nihil amplius boni
aut fpirituofi in Mufculis fit, quod à venis reforptum perq;
ipfas ad cor delatum in eo fermentefcere valeat. Deinde,
idem liquidò cernitur in his quos febres diuturnæ ad fummam maciem redegerunt, qui poftquam edomito morbo
convalefcere cœperint, breviffimo quidem temporis fpatio,
aliment.s inftaurari videmus : denique, in Equis quibufdam
carnis Mufculofæ compages, levi quovis motu facile dif-
folvitur ac eliquatur, paftúque iterum citiffimè reficitur.
Quare ex pulcherrimo hoc cum Nervis quoad Infertionem
confenfu, eodem jure argumentari licet, ipfas Mufculis
movendis infervire, quo, omnes Anatomici à Nervorum
infertionis modo contendunt, ipfos ad eundem ufum à naturâ defignari. De Lymphæ-ductibus autem nihil memo-
ratu dignum ad hanc rem adferre poffum.   Secundo, optimè ab Aquapend : annotatum eft, Tenfionem Mufculi ac
Penis in eo convenire quòd uterque rigefcat : Jam verò Pe-
n.s Tenfionem non ab inflatione aliquâ (quod de Mufculo
etiam Vett: fenfiffe diximus,) provenire, fed à fanguine
per arterias copiofiùs irruente, ibique per materiæ femi-
neæ h. e. laticis Nervofi admiftionem æftuanæ & efferve-
facto, quod nos etiam de Mufculo affirmamus, fatis conftat. Tertiò demonftrat Fallopius (vir, fiquis alius in Con-
fectionibus verfatiffimus) nullam in Toto corpore partem
motu.voluntario fe movere, nifi quæ, præter Fibras, carnem etiam habeat, ac alibi fe explicans, Verum fe carnem
intelligere ait, talémque quæ in Mufculis tomenti modo
fibris impacta eft. Putat enim.verò vigorem movendi, partim à caliditate hujufce carnis accedere, at calor, obfecro,
unde nifi à fanguine, eóque, ut dixi, magis copiofo cùm
pars movetur quam alias? Secus enim perpetuus afforet fe
movendi conatus. Plura autem hac fuper re paulo inferius
                                                    dicentur.

the circulation of the blood continues until nothing good or spirituous remains in the muscles, which, after it [the blood] has been reabsorbed by the veins and returned through them to the heart, could ferment in it. Then the same thing is clearly seen in those whom fevers of long duration have reduced to an extreme meagerness, who, after the disease is over, start to recover. We see that they are restored by food in a very short time. Finally, in some horses the structure of the muscular flesh is easily dissolved and disappears after any rapid movement, but fodder restores it very quickly. Therefore, this very excellent agreement (of the arteries and veins) with the nerves, as far as their insertion is concerned, allows one with the same justification to claim that they serve to move the muscles, since all anatomists assert that because of the way the nerves are inserted, they were designed by nature for the same function. Concerning the lymph ducts, however, I can bring forward nothing worth mentioning on this matter. Secondly, Aquapendente very well remarked that the tension of a muscle and that of the penis agree in that each of them stiffens. Certainly it is sufficiently clear that the tension of the penis does not result from some inflation (which we have said even the ancients understood about a muscle) but from the blood rushing in through the arteries in great abundance, and from its agitation and effervescence by mixing with seminal substance, i.e. nervous fluid, which we assert concerning a muscle. Thirdly, Fallopius (a man very skilled if anyone is in dissections) shows that no part in the entire body moves itself voluntarily unless if, besides fibers, it also has flesh. And, explaining himself elsewhere, he says that by flesh he means that which, in the muscle, is fastened upon the fibers in the way of stuffing. He actually thinks that the strength of moving partly comes from the warmth of this flesh. But, I ask, from where [does] warmth [come] if not from the blood, and in greater abundance, as I have said, when a part is being moved than at other times? Otherwise indeed the effort to move oneself would be continuous. However, more will be said on this matter a little further on.

dicentur.   Quartò, positâ hac causâ, videtur per eam haud
incommodè explicari posse rationem contractionis ac In-
tumescentiæ Musculi: quod sic amplius ostendo.

31. Postquam eâ quam modò loquutus sum sanguinis
eluvione inundatus est totius Musculi habitus, omnesque
ejus pori tam in Carne quàm in Fibris Tendinosis probè
oppleti, non poterit is non valde distendi: Advertendum
est enim tantum sanguinis ex arteria I K O ac tantâ Velo-
citate exire, ut vena socialis L M N ipsi revehendo impar
sit, adeóque in hoc Musculo A B C D sanguinis Circulatio
ad tempus plarè inæqualis reddatur : quod item in Ten-
sione Penis modò dictâ, & cunctis Inflammationibus, Par-
tiumque inde enatis doloribus accidit : hinc Partes sic affectæ duriores ac Tensiores evadunt.   Quò autem hoc me-
lius percipiatur, cogitemus Arteriam I K O è Trunco illo
majori I K O ubi cordis Impulsio est vehementior, sanguinem in minores ramos indeque in habitum Musculi, cum
impetu quodam effundere, unde patet arteriam I K O con-
tinenter impleri: Venam autem L M N eundem, per ramos
exiles ægrè Intrantem, paulatim recipere, ut ex se iterum
versus cor transmittatur, adeóq; hinc perpetuò depleri: Atqui, sanguis in habitum Musculi copiosè effusus, vasa intra
ipsum necessariò comprimet. Ac quoniam major est in arteria I K O à novo hoc sanguinis Impulsu renitentia, quàm in
venâ L M N è quâ lentè admodum ut prius dix , elabitur ,
idcircò istâ compressione Venæ L M N cum suis omnibus
propaginibus, magis quàm Arteria I K O coarctatur: san-
guisq; haud eadem proportione per venam L M N è Musculo exit, quâ ab Arteriâ I K O immittitur. Ut exceam venas
etiam eò comprimi facilius, quòd tunica saltem simplici, in-
trâ Musculos incedant, ut supra ann ot atum fuit. Hinc verò
Musculos A B C D suis membranis conclusus durescere

H 3   ....p..

Fourthly, this cause being admitted, it seems that it can conveniently explain the reason of the contraction and swelling of the muscle, which I thus show more fully in this way.

31. After the fabric of the whole muscle has been flooded by this influx of blood which I have just mentioned, and all its pores in the flesh as well as in the tendinous fibers have been completely filled up, the muscle must be considerably distended. It must indeed be noted that so much blood leaves the artery IKO at such a high velocity that the associated vein LMN is unable to return it, so that the circulation of the blood in that muscle ABCD is completely unbalanced for a while. The same occurs in the erection of the penis, as described above, and in all inflammations and aches in the parts therefrom. Hence, the parts thus affected become harder and tenser. But, to make this better understood, let us imagine the artery IKO originating from the larger trunk IKO, where the impulse of the heart is more violent and the blood flows with some impetus from there into the smaller branches and thence into the fabric of the muscle. It thus appears that the artery IKO is continuously filled. But the vein LMN, penetrating with difficulty through very slender branches, receives the blood little by little to return it to the heart and, therefore, is continuously emptied. But the blood abundantly poured into the fabric of the muscle necessarily compresses the vessels there. Again, since the resistance to this fresh impulse of the blood is greater in the artery IKO than in the vein LMN from which it escapes rather slowly, as I said before, the vein LMN with all its ramifications is more constricted by that compression than the artery IKO. The blood does not leave the muscle through the vein LMN in the same proportion as that at which it is sent in by the artery IKO. Moreover, the veins are the more easily compressed because within the muscles they have only a single tunic, as was mentioned above. Hence, the muscle ABCD enclosed in its membranes begins to harden and to swell

28

incipit, ac ad latera R. Q. tumere, secundùm autem longitudinem A C contrahi. Cúmque Caput A prorfùs immobile fit, Tendinis extremum C versùs A adducitur.

32. Accedo jam ad demonstrandum, Hujusmodi Intu-
mescentiam musculi, quamvis exigua fingatur, non tantum satis valere ad quodlibet corporis membrum attol-
lendum, sed etiam ad aliud quodcunque Pondus Tendini
in C, appensum, quod ab ipsius capite A, ac fibris per ipsum
decurrentibus sustineri possit: quantum autem illud sit,
conjecturâ saltem ab experimento Sect. 20. assequi licebit.
Quare esto in Fig. 2. secundus Cubitum flectentium
Musculus (Brachiæus dictus) M E O, Principium ejusdem in esse humeri M, venter E, Tendo autem, quo lato
quidem ac Carnoso in Ligamentum articuli, & cubiti ra-
diique Appendicem inseritur, sit O. Pondus movendum
(adjuvante etiam bicipite de quo eadem est ratio) esto
tota illa Ossium carniúinque compages ab osse Cubiti
T, ac Radio V, usque ad extremos digitos. Quoniam autem,
omnis harum partium moles circa articulationem ossium
Humeri cùm ossibus cubiti radiíque tanquam centrum ro-
tatur, ideò musculus hic incommodam admodùm In-
sertionem nactus videtur, centro utpote nimis propin-
quam. Illud tamen non obstat, quò minùs ex principiis
nostris clarè intelligatur, quemadmodum hoc totum pondus, etsi vel centies esset gravius, facilè moveri posset.
Accipiamus ergò in ventre Musculi punctum aliquod, pu-
tà E, dumque Musculus tumescit usque in M D O, cogitemus hoc punctum E progressum esse in D, cubitumque
proinde parùm inflexisse. Tum facto in ipso Articulo O
Centro, Semidiametro O E ducatur Circulus E I N, Tan-
git ipsam in puncto E recta E D, & ad punctum in Tangente D continuetur Radius O q, ac loco curvæ hæc

O A D

towards the sides R.Q. but to be contracted over its length AC. Since the head A is completely immobile, the extremity C of the tendon is brought towards A.

32. I now come to the demonstration that this sort of swelling of the muscle, however small it may be thought to be, is not only sufficient to raise any limb of the body but also any other weight suspended from the tendon at C, which can be supported by the head of the muscle A and the fibers running through the muscle. How great this weight may be can be conjectured from the experiment in Sect.20. In Fig.2, let MEO be the second muscle among the flexors of the elbow (called brachialis), M its origin from the bone of the humerus, E its belly, and O the tendon, broad and also fleshy, by which it is inserted into the ligament of the joint and the tuberosity of the ulna and radius.[8] Let the weight to be moved (with the help also of the biceps for which the explanation is the same) be all that combined structure of bones and flesh from the ulna T and the radius V to the tips of the fingers. Since, however, all the mass of these parts rotates about the articulation of the bones of the humerus with the bones of the ulna and the radius, as if it were a pivot, this muscle seems to be provided with an extremely inconvenient insertion in so far as this is too close to the center of rotation. However this does not prevent that, for it is clearly understood from our principles how all this weight, and even if it were hundred times heavier, could be easily moved. Let us therefore suppose some point E in the belly of the muscle. While the muscle swells out to MDO, let us imagine that this point E has proceeded to D and, henceforth, that the elbow has accordingly flexed a little. Then the center O having been made in the joint, and the radius OE, let there be drawn a circle EIN. Let a straight line ED be drawn tangent to the circle at E, and the radius Oq be continued to the point D on the tangent. In place of the curved line

---

[8]The fibers of the brachialis "converge to a thick, broad tendon attached to the ulnar tuberosity and a rough impression on the anterior aspect of the coronoid process. It may be divided into two or more parts, fused with brachioradialis, pronator teres or biceps, or send a tendinous slip to the radius or bicipital aponeurosis." (*Gray's Anatomy*, [Churchill Livingstone, Edinburgh, etc, 1989], p.615)

39

O A D intelligamus rectam O q D. Erit itaque spatium à puncto E, dum Musculus tumescebat transmissum recta E D. Pars autem secantis extra ambitum Circuli q D ostendit spatium à cubito propè centrum motûs peractam dum introrsum flectebatur. Jam concipiamus Vim illam quâ E versùs D movetur, vicem cantilli cujusdam ponderis habere : Dico, semper majorem futuram proportionem Velocitatis Motûs puncti E ad Velocitatem Ponderis attollendi, quàm est gravitatis ejusdem ponderis (quam maxima cunq; sit) ad gravitatem vel minimi ponderis in E trahentis ab E versùs D, cujus locum vim intra Musculum agentem obtinere modò supposui; unde consequitur Musculum M N O extumescentem facilè pondus Cubiti attrahere, quod ità ostenditur.

Esto Pondus Cubiti ·   a:
Vis movens in E·   b.
    Sitq;

$$\frac{a}{b} = \frac{c}{d}$$

Sumatur e multò minor quàm d   Ergò $\left.\begin{array}{c} \frac{c}{e} \end{array}\right\}$ $\frac{c}{d}$

Sint autem   c,   e,   f. $\div$

Fiat quoq; $\frac{c}{f} = \frac{ID}{qD}$

Sed per 35 e. 3.   ID. ED. qD $\div$

Ergò $\frac{ED}{qD} = \frac{c}{f} = \frac{c}{e} \quad \frac{c}{d} = \frac{a}{b}$

33. A duabus hisce caussis omnem Musculorum Motum primario inchoari opinor: Totaque vis illa ac momentum, quo singuli Musculi suos Antagonistas superant, ab iis pendet. Ac de Nervorum quidem efficaciâ res ferè extrà dubium est, ut nullus adeò ambigat quin ipsi principem locum in Motu Musculi inchoando obtineant, quia

OAD, let us consider a straight line OqD. Thus, while the muscle was swelling, the distance covered by point E was the straight line ED. But the part of the secant outside the periphery of the circle qD indicates the distance traveled by the ulna near the center of movement while being flexed inwards. Now let us conceive that the force by which E is moved towards D plays the role of a very small weight. I claim that the ratio of the velocity of the movement of point E to the velocity of the weight to be raised is always greater than the ratio of the gravity of this weight (however great it may be) to the gravity of even the smallest weight in E dragging from E towards D which I supposed just before to replace the force acting within the muscle. Consequently the muscle MNO, when swelling out, easily pulls the weight of the forearm, which is thus demonstrated.

Let the weight of the forearm be	a.
Moving force at E	b.
And let	$a/b = c/d$
e is supposed to be much less	
than d. Thus	$c/e >> c/d$
But let	$c/e = e/f$
And let	$e/f = ID/qD$
But, from 35.c.3	$ID/ED = ED/qD$
Thus	$ED/qD = e/f = c/e >> c/d = a/b$

33. I think that all movement of muscles primarily starts from these two causes. All the force and moment by which the different muscles overcome their antagonists results from them. And the efficacy of the nerves is a matter quite beyond doubt so that, so far, nobody disputes that they play the main role in initiating the movement of the muscle since,

30

quia his læfis omnis hujus machinæ ufus tollitur. Meritò
autem quæri poteft, Annon ifta vis five liquor è Nervis ex-
filiens, iftuifmodi in fe acrimoniam habeat, licet minùs vehe-
mentem, quàm in liquoribus Ptarmicis videmus: cujus beneficio, in Tendinofum Mufculi caput exftillans omnes ejus
fibras ad fe ftatim contrahendum velut aculeis quibufdam
extimu'et; ut cùm fternutatio à pharmaco excitatur, aut
uti materia acris intra nares retenta, à folis luce rarefit,
hominémq; fternuere facit. Neq; aliter Medicamenta pur-
gantia, Vomitúmq; provocantia operari videntur: dum
enim ventriculus ac inteftina earum virulentiâ irritantur,
fibras fuas carneas valicè conftringendo id quod noxium
& ingratum fentiunt, magno conatu ejiciunt: ac memini
fanè leviffimo attactu liquoris cujufdam peracris, quem ab
Honoratiffimo viro noftrique Temporis ornamento maxi-
mo D. Boyl acceperam, Mufculum ex humano femore ex-
cifum, momento fe contrahere: ex quibus confici videtur,
Laticem hunc Nervofum cùm folis Fibris Tendinofis, mo-
vendo Mufculo fatis effe. Equidem haud inficior, nonnihil
fortè hinc ad Fibrarum contractionem accedere, fed non
antequam Priores illæ caufæ ità eas adjutârint, ut poffint
æquipollentem fe contrahendi vim quæ in Mufculo antagoniftâ eft, exfuperare: Huic itaq; Objectioni non unâ ratione obviam ire licebit: Namque primò, fi non ità fit
quod modò dixi, nondum fanè apparet, quo pacto Fibrarum virtus illa fe contrahendi in altero Mufculo, ex in-
fluxu laticis Nervofi, ità viribus augeatur, ut Mufculum
fuum contrahere poffit, dum interim incolumis manet
æqualis ipfi virtus in Mufculo Antagoniftâ, quod profectò
id ipfum eft de quo maximè Quæftio inftituitur, ut Sectione 7. fubinnui. Secundò, Annotavit excellens Anatom.
Fallop. Omnem corporis partem, quæ motus voluntarii

caput

if the nerves have been injured, all function of this machine is abolished. It could be deservedly asked whether that force or liquor springing out from the nerves has in it a pungency, albeit less intense, of the sort we see in sneezing fluids. By means of this, when trickling out into the tendinous head of the muscle, it excites, as if with stings of some sort, all the fibers to draw themselves together immediately, as when sneezing is provoked by a drug, or as the loose fleshy substance within the nose is rarefied by the light of the sun and makes a man sneeze. Neither do the medicines provoking purging and vomiting seem to work otherwise: while indeed the stomach and bowels are irritated by the virulence of these substances, they expel with a great effort what they feel to be noxious and unpleasant by strongly constricting their fleshy fibers. I certainly remember that, at the lightest touch of some very pungent liquor which I had received from a very honourable man and the greatest ornament of our times, Mr Boyle, a muscle removed from a human thigh contracted instantly. These facts seem to show that this nervous fluid, together with only the tendinous fibers, is sufficient to contract a muscle. I do not deny that perhaps something of this contributes to the contraction of the fibers but not before those prior causes have so assisted them that they can overcome the equivalent force of contraction which is present in the antagonistic muscle. Thus it will be possible to meet that objection with not one reason alone. Firstly, indeed, if it is not as I said, it is not yet sufficiently clear by what means that ability of the fibers in one muscle to contract themselves, may be so increased in strength after an influx of nervous fluid that it can contract the muscle while an equal ability in the antagonistic muscle remains nevertheless unimpaired. This is the very thing with which the investigation is concerned, as I have suggested in Sect.7. Secondly, the excellent anatomist Fallopius has noted that every part of the body which is capable of voluntary movement

capax eſt, fibris donatum eſſe: atqui iſtas fibras una cum
ſuo Nervo ad motum concipiendum non ſufficere, idem
ille oſtendit: " Eſt (inquit) neceſſaria propenſio atque
" apta natura ſubjecti corporis ad concipiendam hanc mo-
" vendi facultatem, quæ quidem proper ſi à carne de-
" pendet. Quoniam video corpora quædam fibroſa cum
" adjunctis nervis à quibus facultatem habent communi-
" catam,& tamen ſeipſa non movent. Et paulò poſt, ' Hoc
" idem pluribus exemplis confirmare poſſem, ſed unum hoc
" mihi ſatisfacit, Omnes Membranas quæ ſeipſas movent
" carneas eſſe factas: Si dicant eum eſſe carnis Impactæ
uſum, ut ſint tanquam ὑπομόχλιον, ſeu potiùs Trochlea per
quam Fibræ veluti funes chordæquè ducantur : optimè
hanc excuſationem amolitur hæc Fallopii obſervatio:
quippe , ſi fortè id eſſet aliquid in Muſculis artus mo-
ventibus, tamen ad membranas contrahendas haud opus
eſt: namque hæ cùm ex infinitâ fibrarum contextuâ fi-
ant, non aliter quàm fibra aliqua ſingularis aut Corium validè extenſum, contrahuntur : ad quod, nihil certè quod
Hypomochlii Trochlævè vicem habeat requiri, ſatis notum
eſt.   Cum itaque in Panniculo carnoſo omnes Partes fibro-
ſæ ſint, omneſque tranſmiſſam habeant per nervos facultatem, ac etiam ſentiant; Cur non omnes etiam moventur? Sed eæ tantùm quæ carnem habent ſib is Incruſta-
tam. Quibus adjicere quoque poſſem, Iſtos Muſculos quæ
meminit V.Cl Deuſing. in eruditâ ſuâ de motu Muſc. Ex-
ercitatione :' Sunt (inquit) Muſculi qui una ſecum movent
' ſolam cutim, ut labiorum, faciei, &c. vel aliud quid leve
' dentaxat commovent, ut Muſculi linguæ laryngis, &c.
' talium fibræ vix in viſibiles tendines coaleſcunt. Equidem
omnibus rite perpenſis, opinor magis fabricæ Muſculi congruere, ut ejus movendi ratio è principiis iſtis mechanicis

q. 2

is endowed with fibers. He has also shown that these fibers together with their nerve are not sufficient to generate the movement. He says: "There is a necessary inclination and a suitable nature of the body of the subject appropriate to the conception of this faculty of moving. This inclination depends on flesh since I see some fibrous bodies with connected nerves, by which they have the faculty imparted, and yet they do not move themselves." And a little later: "I could confirm this same thing with several examples but this one satisfies me: all the membranes which move themselves are made of flesh." If people say that the function of compacted flesh is as if it were a fulcrum, or rather a pulley through which the fibers are led like ropes or cords, this observation of Fallopius very well silences this objection. Obviously, if this were perhaps something of some use in the muscles which move joints, it is scarcely needed in contracting the membranes, since these are made up of an infinite weaving together of fibers, and are contracting no differently from any single fibre or skin that is very tautly stretched out. To this end it is sufficiently known that certainly nothing is required which has the function of a fulcrum or a pulley. Since all parts of fleshy coverings are fibrous and since all have a faculty transmitted through the nerves, and even feel, why are they not also moved? But only those which have flesh encrusted with fibers do it. To these [arguments] I could also add those muscles which the famous man Deusing mentions in his learned treatise on the movement of muscles. He says: "There are muscles which move together with themselves only the skin, like those of the lips, of the face, etc. or move something which is only light like the muscles of the tongue, the larynx, etc., and the fibers of such muscles hardly coalesce into visible tendons." All things having been carefully considered, I myself am of the opinion that it is more in accord with the structure of the muscle to deduce that its movement results from those mechanical principles

32

quæ Sect. 3?. allata sunt, quàm ex aliis quibuscunque dedu-
catur, adeóque is sit Carnis Infarctæ usus, quem superius adstrux'. Attamen illa, ad particulares quorundam
Musculorum motus demonstrandos respectu peculiaris cujusdam quem obtinent situs, ut à Doctiss. Charletono ali-
isque factum est, rectè adhiberi non negem.

34. Ex dictis modò consequitur, haud tantùm à Nervis,
verùm etiam à sanguine in carnem Musculosam copiosiùs
per arterias illato omnem ejus motum inchoari: His autem jam tertia illa succedit, quæ à spontaneâ fibrarum
contractione proficiscitur : Quo enim magis intumescit
Musculus versus Q & R, eò major ipsis conceditur se contrahendi facultas; Accidit autem interdum, tertiam hanc
causam ab externâ Vi adjutam agere per se, ut in casu Sect.
6. cùm alter alteri subito brachium flectit. Imò multum
ad hoc facit, sanguinis advenientis calor; ut è Musculo cocto
apparet, qui vix quartam pristinæ longitudinis partem ser-
vat : Taceo aculeatam fortè vim qualem Sect. 33. diximus
in latice hoc nervoso inesse, unde fibræ nonnihil irritentur,
certè ex attactu liquoris D. Boyl haud parùm se contraxit
Musculus, nempe quia Antagonistâ caruit. Præterea
denique, ipsam illam humoris copiam quo Musculus tur-
gescit aliquantu'ùm hanc contractionem adjuvare, eodem
ritu quo ille Venetus immani magnitudine columnas, fu-
nibus aquâ aspersis dimovit. Atque ita omnibus his causis
valicè simul agentibus, Musculi finis versus originem
pertrahitur.

35. Hæc autem ægrè ab iis assensum impertrabunt, qui-
rum animus ista de spirituum influxione occupavit opinio;
quippe vix crediderint velocissimas illas, quas videmus,
Membrorum agitationes ac citissimas Tensionum ac Re-
laxationum alternationes in Musculis Antagonistis, hanc
                                                    providere

which have been presented in Sect.32, than from any other principles whatsoever. Therefore, this function of the stuffed-in flesh is that which I explained above. However, I would not deny that these principles are correctly applied to the demonstration of particular movements of certain muscles with respect to the peculiar situation which they obtain, as the well-learned Charleton and others have done.

34. From what was said it results that every movement of a muscle is initiated not only by the nerves but also by the blood more copiously carried to the muscular flesh through the arteries. But these causes are now followed by a third one which arises from the spontaneous contraction of the fibers. Indeed the more the muscle swells towards Q and R (Fig.1), the more the fibers acquire the faculty of contracting. But it happens sometimes that this third cause helped by an external force acts by itself as in the instance of Sect.6 when one suddenly bends the arm of somebody else. Assuredly, the warmth of the arriving blood does much for this as is apparent from a boiled muscle which occupies hardly one quarter of its previous length. I do not mention that perhaps the sting-like force which we claimed in Sect.33 is present in this nervous fluid, whereby the fibers are considerably irritated. Assuredly as the result of a touch of Mr Boyle's liquor the muscle contracted quite considerably, no doubt because it lacked an antagonist. Finally, I mention in passing, that the great quantity of moisture by which the muscle is made to swell contributes somewhat to this contraction in the same way that that Venetian put in motion columns of enormous size with ropes on which water was poured.[9] Thus, as a result of all these causes acting together, the end of the muscle is pulled towards its origin.

35. But these explanations will scarcely gain access to those whose minds are occupied by that opinion concerning inflation on account of the spirits, since they will hardly believe that these very quick agitations of the limbs, which we observe, and the extremely rapid alternations of tensions and relaxations in antagonistic muscles, can

---

[9] See Introduction, Footnote 45, p. 15.

provenire poſſe. At, inquam, Hæc omnia etiam momento temporis ab hiſce cauſis præſtari: Simul enim ac Nervi Fibrillæ H E G ad cerebrum protenſæ concutiuntur, continuò iſtæ liquoris guttulæ ex omnibus ejus propaginibus excidunt, ut embolo ſyringis vel leviſſimè intruſo, liquor ſtatim exſilit : namq; ſumma intelligi debet totius corporis à ſuccis ſpirituoſis ac vitalibus plenitudo. Ut verò aqua oleo vitrioli affuſa, ità liquor hic in ſanguine, momento efferveſcentiam concipit : Eodémq; plane inſtanti, ſanguis per arteriam I K O quaſi aqua è ſiphone aperto Epiſtomio effluit. Ad extremum non citius tumere incipit Muſculos, quin ſe etiam contrahant fibræ; adeóq; cuncta ſimul ac ſemel, ictu quaſi oculi, contingant. Ea autem quam dixi efferveſcentia admodùm repentè ceſſat, ſpiritúſque actuoſi momento ferè per membranas Muſculi diſſipantur ; ac, niſi novus ſtatim acceſſerit, ſublatâ circulationis ſanguinis inæqualitate per fibrarum antagoniſtæ contractionem Muſculus ſtatim retrahitur & flacceſcit, ſanguiſq; illicò majori copiâ per venam L M N elabitur; hinc etiam exercitia nimis violenta ſpiritus abſumunt, eóſq; per ſudores erumpere faciunt. Quinetiam membra poſt vehementiores corporis motus (præſertim in his qui iis non ſunt aſſueti) valdè dolent ac languent. Quàm citò autem effluat ſanguis iterúmq; refluat, ſubita illa ex cogitationibus venereis Penis erectio, ejuſdémq; ab exitu Seminis flacciditas, Totius faciei ex pudore ſuffuſio, momentanea quaſi in conſternatione ex omni corporis parte ejuſdem ad interiora receſſio, aliáq; hujuſmodi docent. Adhæc, quò minor eſt attollendi ponderis ad vim moventem proportio, eò citiùs conficitur ſpatium E D. & q D. Hinc magna pondera veſicæ appenſa ejuſdem inflatione, aut aquæ in eandem immiſſione, ſatis celereter elevantur. Deinde, dum pars brachii,

B

result from them. But I say that all these things are carried out by these causes even in an instant. Indeed, at the same time that these fibrils of the nerve HEG, extended to the brain, are struck, these droplets of liquor exude continuously from all its branchlets, just as liquor spurts out at once when the piston of a syringe is pushed in very lightly, for it must be understood that the whole body is filled completely with spirituous and vital juices. Like water added to oil of vitriol, so this liquor instantly rouses an effervescence in the blood. In the same instant blood flows through the artery IKO like water from an opened pipe. No sooner does the muscle begin to swell at its boundary than the fibers are also contracted, and so everything occurs at once as if in the twinkling of an eye. However, that effervescence which I have mentioned ceases very rapidly and the very active spirits are dissipated through the membranes of the muscle almost in an instant and, unless a new impulse arrives at once, the muscle is immediately pulled back and made flaccid when the imbalance of the circulation of the blood ceases due to the contraction of the fibers of the antagonist. And the blood immediately flows out in greater quantity through the vein LMN. Hence, also, excessively violent activity exhausts the spirits and makes them break out through sweats. This is also the reason why the limbs are very painful and weak after more violent movements of the body (especially in those who are not accustomed to them). The sudden erection of the penis resulting from sexual thoughts and its flaccidity after the emission of semen, the sudden flushing of the whole face from shame, the instantaneous recession of the blood from every part of the body to the inside in fear, all show how rapidly the blood flows in and flows out again. Moreover the smaller the ratio of the sustained weight to the moving force, the more rapidly the distances ED and qD are travelled. Hence, heavy weights attached to a bladder are fairly quickly raised by inflating the bladder with air or with water. Then, one must think that, while the part of the arm

chii, quæ articulationi vel contro propinqua est, transit per illud spatium minus q D, alterum ejus extremum, spatium longè majus eodem tempore emetiri putandum est. Denique siquis attentè consideret, istos quos tantoperè crepant spiritus, nusquam extra sanguinem & laticem Nervosum in corpore reperiri, atq; adeò reipsa Liquores esse; eò forsan animus à veritate aliena quæ diximus existimabit.

F I N I S.

which is close to the joint or to the centre of rotation travels this smaller distance qD, the other extremity traverses a much longer distance in the same time. Finally, if one considers carefully that those spirits about which people make so much noise are found nowhere in the body outside the blood and the nervous fluid, and are in fact liquids, he would judge that what we have said is perhaps the less inconsistent with the truth.

THE END

INDEX